D0982996

Fractals for the Classroom:
Strategic Activities Volume Three

Springer
New York
Berlin
Heidelberg
Barcelona
Hong Kong
London
Milan
Paris
Singapore
Tokyo

Heinz-Otto Peitgen
Hartmut Jürgens
Dietmar Saupe
Evan Maletsky
Terry Perciante

Fractals for the Classroom: Strategic Activities Volume Three

National Council of
Teachers of Mathematics

Springer

Heinz-Otto Peitgen
CeVis and MeVis
University of Bremen
Universitätsallee 29
28359 Bremen
Germany *and*
Department of Mathematics
Florida Atlantic University
Boca Raton, FL 33432 USA

Hartmut Jürgens
CeVis and MeVis
University of Bremen
Universitätsallee 29
28359 Bremen
Germany

Dietmar Saupe
Department of Computer Science
University of Leipzig
PF 920, 04009 Leipzig
Germany

Evan Maletsky
Department of Mathematics and
Computer Science
Montclair State University
Upper Montclair, NJ 07043 USA

Terry Perciante
Department of Mathematics
Wheaton College
Wheaton, IL 60187-5593 USA

Published in cooperation with the National Council of Teachers of Mathematics (NCTM).

Camera-ready copy supplied by the authors.
Printed and bound by Maple-Vail Book Manufacturing Group, Binghamton, NY.
Printed in the United States of America.

9 8 7 6 5 4 3 2 1

ISBN 0-387-98420-8 Springer-Verlag New York Berlin Heidelberg SPIN 10424840

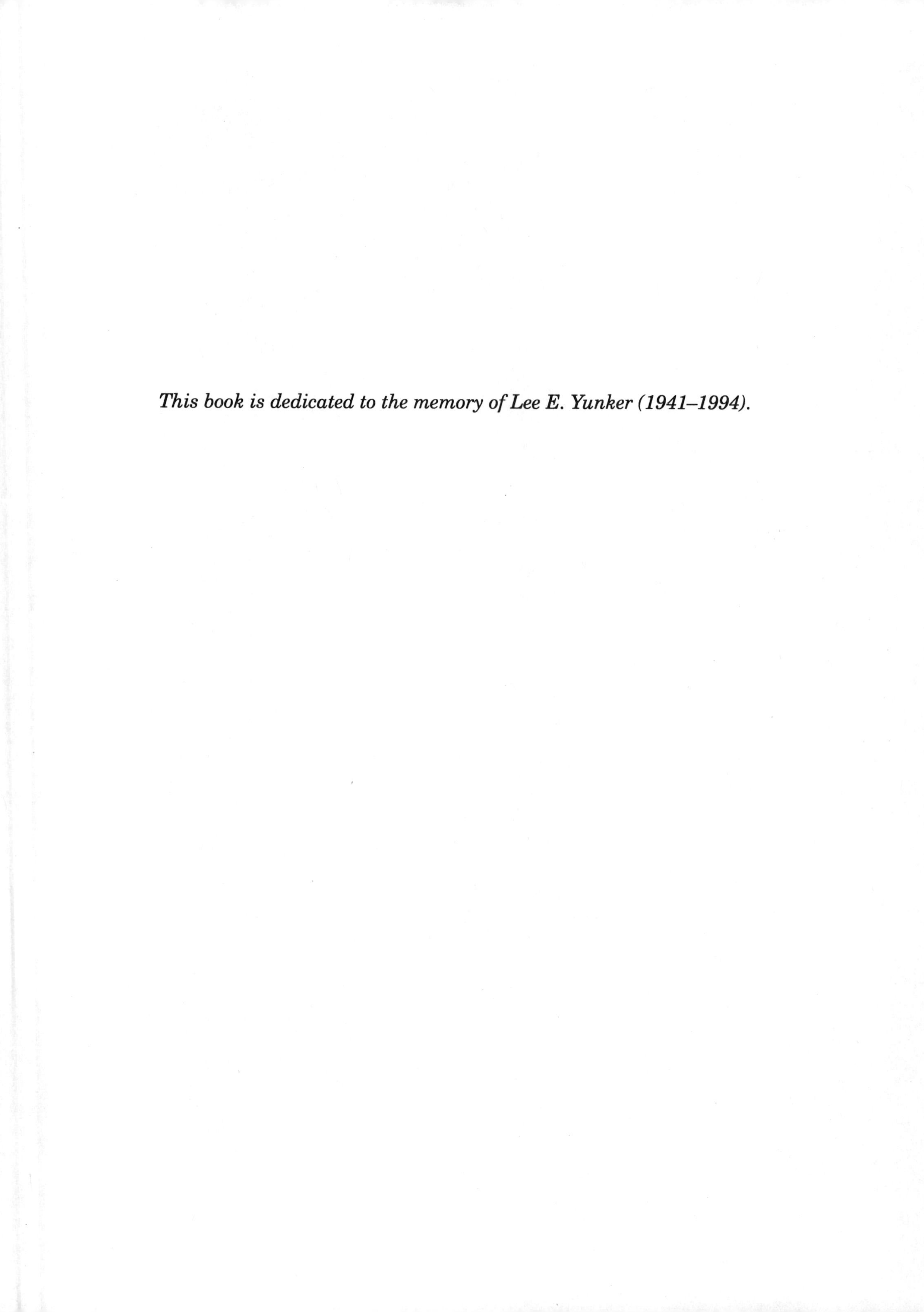

This book is dedicated to the memory of Lee E. Yunker (1941–1994).

Preface

This third and final volume of *Strategic Activities* on fractal geometry and chaos theory focuses upon wonderful images that for many people have provided a compelling lure into an investigation of the intricate properties embedded within these images. By themselves the figures possess fascinating features, but the mechanisms by which they are formed also highlight significant approaches to modeling natural processes and phenomena. The general pattern and specific steps used to construct a fractal image illustrated throughout this volume together comprise an iterated function system. The objective of the following units will be to investigate the processes and often surprising results of applying such systems.

Individuals willing to traverse the path of activities that follow will quickly find the same remarkable integration between geometry and numeration, theory and practice, and mathematics and technology that has characterized the prior volumes. All of these connect to scientific applications in a way that suggests the power and importance of these new methods.

The contents of this volume joined with the details contained in the prior two books provide more than a survey of the significant aspects of fractal geometry and chaos theory. In addition to the key notions of this new and emerging discipline, the dynamic nature of the research and the experimental characteristics of related applications together provide a paradigm for classroom activity that engages hands and minds, individuals and groups.

Ultimately, it is our hope that the collection of activities and commentary represented by this volume and the preceding two editions will in some small way add impetus to developing a new generation of mathematicians and scientists who will exercise vigorous inquiry, experimental curiousity, and insightful theory for a new millennium.

This volume concludes our series of *Strategic Activities* on fractal geometry. We are very grateful to Jana Wallace from San Juan College in Farmington, New Mexico, for providing her personal view of how she became involved and interested in using fractals and chaos in her classroom as the foreword of this booklet. We also thank Torsten Cordes, who again took great care in producing the fine figures for this volumes, and Raimund Albers for proofreading and help with the calculator codes. Of course, all of the remaining errors have to be blamed upon us, the authors.

Evan M. Maletsky
Terry Perciante

USA, 1998

Heinz-Otto Peitgen
Dietmar Saupe
Hartmut Jürgens

Germany, 1998

Authors

Hartmut Jürgens. ∗1955 in Bremen (Germany). Dr. rer. nat. 1983 at the University of Bremen. Research in dynamical systems, mathematical computer graphics and experimental mathematics. Employment in the computer industry 1984–85, since 1985 Director of the Dynamical Systems Graphics Laboratory at the University of Bremen. Author and editor of several publications on chaos and fractals.

Evan M. Maletsky. ∗1932 in Pompton Lakes, New Jersey (USA). Ph. D. in Mathematics Education, New York University, 1961. Professor of Mathematics, Montclair State University, Upper Montclair, New Jersey, since 1957. Author, editor, and lecturer on mathematics curriculum and materials for the elementary and secondary school, with special interest in geometry. First Sokol Faculty Fellow Award, Montclair State, 1991, and 1993 Montclair State Distinguished Teacher Award. 1991 Outstanding Mathematics Educator Award, AMTNJ.

Heinz-Otto Peitgen. ∗1945 in Bruch (Germany). Dr. rer. nat. 1973, Habilitation 1976, both from the University of Bonn. Research on nonlinear analysis and dynamical systems. 1977 Professor of Mathematics at the University of Bremen and from 1985 to 1991 also Professor of Mathematics at the University of California at Santa Cruz. Since 1991 also Professor of Mathematics at the Florida Atlantic University in Boca Raton. Visiting Professor in Belgium, Italy, Mexico and USA. Author and editor of several publications on chaos and fractals.

Terence H. Perciante. ∗1945 in Vancouver (Canada). Ed. D. in Mathematics Education, State University of New York at Buffalo, 1972. Professor of Mathematics at Wheaton College, Illinois, since 1972. Received a 1989–90 award for Teaching Excellence and Campus Leadership from the Foundation for Independent Higher Education. 1977 and 1994 Wheaton College Teacher of the Year. Author of several books, study guides and articles at the college level.

Dietmar Saupe. ∗1954 in Bremen (Germany). Dr. rer. nat. 1982, Habilitation 1993, both from the University of Bremen. Visiting Assistant Professor of Mathematics at the University of California at Santa Cruz, 1985–87, Assistant Professor at the University of Bremen, 1987–93, Professor of Computer Science at the Albert-Ludwigs-University of Freiburg, 1993-1998, and at the University of Leipzig since 1998. Research in image processing, computer graphics, visualization, experimental mathematics, and dynamical systems. Author and editor of several publications on chaos and fractals.

Table of Contents

Connections to the Curriculum

Students can best see the power and beauty of mathematics when they view it as an integrated whole. These charts show the many connections these strategic activities have to established topics in the contemporary mathematics program.

The National Council of Teachers of Mathematics, in its *Curriculum and Evaluation Standards for School Mathematics*, stresses the importance of mathematical connections:

> *"The mathematics curriculum should include investigation of the connections and interplay among various mathematical topics and their applications so that all students can*
>
> - *recognize equivalent representations of the same concept;*
> - *relate procedures in one representation to procedures in an equivalent representation;*
> - *use and value the connections among mathematical topics;*
> - *use and value the connections between mathematics and other disciplines."*

Unit 7

Connections	7.1	7.2	7.3	7.4	7.5	7.6	7.7	7.8	7.9
Function Composition		●	●					●	
Mappings		●	●				●	●	●
Geometric Transformations	●	●					●	●	●
Geometric Convergence	●		●					●	●
Visualization	●	●	●	●	●			●	●
Counting and Combinations		●		●	●	●			
Iteration and Recursion	●	●	●	●	●				
Pascal's Triangle				●	●	●			
Binomial Expansions					●				
Number Patterns	●	●	●	●	●	●			
Geometric Patterns				●	●	●			
Binary Numbers			●			●			
Limits	●	●	●						
Graphing Calculator								●	

Unit 8

Connections	8.1	8.2	8.3	8.4	8.5	8.6	8.7	8.8	8.9
Linear Equations			●				●		
Function Composition		●	●						●
Linear Transformations			●	●			●		
Geometric Transformations	●	●	●	●	●	●	●	●	●
Mappings	●	●	●	●	●	●	●	●	●
Visualization	●	●	●	●	●	●	●	●	●
Geometric Convergence				●	●	●	●	●	●
Similarity and Symmetry	●	●	●	●	●	●	●	●	●
Counting Techniques		●			●			●	
Iteration					●	●	●	●	
Groups		●						●	●
Matrices			●						
Coordinate Geometry			●						
Graphing Calculator							●		

Foreword

The first time I heard the word fractal, I was walking along a street in New Orleans. I was attending the l991 national NCTM Conference. The workshops and sessions I attended had my head spinning with new ideas for my junior high school math classes. The hundreds of materials at the exhibits filled me with awe. In fact, when I left the exhibit hall, my MasterCard was sizzling, and I could barely manage to carry my purchases. Therefore, I was anxious to get back to my hotel room to unload my packages. As luck would have it, I found myself behind a large group of teachers who were walking very slowly while animatedly discussing the topic of an upcoming large-group lecture session with Heinz-Otto Peitgen as the speaker. The topic? Fractals! Since I was forced to either follow the teachers at a very slow-moving pace or dart into the street into oncoming traffic to try to get around the group, I chose the more safe and sensible course and elected to follow and eavesdrop on their conversation. I gleaned the information that fractals were beautiful geometric shapes new to modern mathematics. I heard such terms as "Mandelbrot Set," "Lorenz Attractor," and "Chaos Theory." While the topic sounded very interesting, I discounted the idea of even attending the session, because I thought it would have no application nor capture any interest in the world of junior high school mathematics. At that time I had no idea how very wrong I was. I regretted my decision to not attend Heinz-Otto's presentation the very next day, though, when I overheard wonderful comments about the session and the "buzzword" of the conference seemed to be "fractals." Ah well.. perhaps I would have another opportunity again someday. In the meantime, I began to question my colleagues, and, though I was not convinced at that time that younger students could understand the concepts, I became very intrigued personally.

My "someday" came again at the 1992 national NCTM Conference in Nashville, Tennessee. My fascination with fractals had grown considerably over the past years, so I enthusiastically attended a workshop offered by Evan Maletsky on Wednesday morning which concentrated on fractals in the middle school, just my area! On Thursday I attended a workshop presented by Lee Yunker in the morning and another presented by Terry Perciante that afternoon. In the meantime, I fried my credit card again buying everything written by Peitgen, Maletsky, Perciante, Yunker, Saupe, and Jurgens that was available at either the Springer-Verlag booth or the NCTM booth. A good friend introduced me to Heinz-Otto, Evan, Terry, and Lee during a break and found me a seat that evening on the front row of the huge lecture hall in which Heinz-Otto and Benoit Mandelbrot were to present their lecture, "Fractals for the Classroom: The Fascinating Concept of Chaos and Fractals." In fact, I was seated between these two illustrious gentlemen! All of the workshops and the evening lecture filled me with a tremendous enthusiasm, to not only learn more about this new branch of mathematics to satisfy my own interest, but to share it with my students. That was my "fractal conference," because I spent most of my time learning everything I could about the subject. The men I heard emanated an enthusiasm and passion for the subject that inspired me. I was rapidly realizing that fractals are not just beautiful geometric pictures, but that they are representations of relatively simple yet extremely powerful mathematical concepts, concepts that are well on the way to explaining and predicting behavior of many dynamical or nonlinear systems such as regulation of heartbeats, fluid turbulence, population growth, propagation of genes, and weather cycles, to name just a few. Besides being awed by the sheer beauty and power of this new branch of mathematics, I was overwhelmed by the interest these men had in the teaching of these concepts. The men I listened to at that conference, Evan, Terry, Lee, and Heinz-Otto, are not just researchers and authors, but are outstanding educators as well. They are deeply committed to the development of fractal learning activities and strategies that can be incorporated into the mathematics curriculum at every level.

At the Nashville conference I found a flyer announcing a Fractals, Chaos, and Dynamics Symposium at Fermi National Accelerator Laboratories in Batavia, Illinois. I was off again for a week of fractals in the summer of 1992. Heinz-Otto, Terry, Evan, Lee, Dietmar Saupe, and Hartmut Jurgens conducted the thrilling, invigorating, information-packed, and exhausting week of fractals, chaos, and dynamics education. That year I had been awarded a GTE GIFT Foundation grant to develop, with the help of a colleague, a course called Creative Project Development. I included numerous lectures, activities, and discussions concerning fractals and found, to my delight, that my students not only understood the concepts but enjoyed the lessons and went far beyond my expectations of their abilities. I discovered a flexibility and openness in their young minds that is sometimes missing in adults. They loved the Chaos Game played with dice; they marveled at the development of Sierpinski's Triangle; they understood the concept of self-similarity, infinite complexity, endlessly repeating, and ever diminishing; they learned how to calculate fractal dimension and find "addresses" of triangles in Sierpinski's Triangle; they balked at the idea that broccoli romanesco — a vegetable (ugh!) — exhibits a beautiful fractal structure; and, as junior high students, they began to understand the ideas of limits and infinity when examining the perimeter and area of various stages of Sierpinski's Triangle.

Currently, I am teaching at San Juan College where some of my courseload includes teaching math in the elementary teacher training program. They are enthralled with fractal geometry and, together, we have discovered many ways to teach the concepts to even very young children. For instance, we have demonstrated sensitive dependence upon initial conditions as an underlying precept of chaos theory by giving each child a dandelion stem with the fluff instead of the flower on it, having each child wave his hand in front of it in a different direction and then observing the ways in which the fluff scatters. We have emphasized fractal structure in nature, and they delight in discovering fractal patterns for themselves. Young children are also very capable of following a coloring scheme in Pascal's Triangle in order to produce a likeness of Sierpinski's Triangle. They can even extend the coloring pattern without the benefit of numbers to follow, because they can visualize "how the little triangles should fit together."

One of my greatest delights has been in watching my daughter, Kelley, develop a tremendous interest in fractal geometry due to my interest in the subject. As a seventh-grader, she participated in all the fractal workshops presented by our friends at the Seattle NCTM national conference. She paid close attention to the material presented by Mitchell Feigenbaum and Heinz-Otto at their evening lecture session, even though I doubt, at that age, she understood much of what was being said. The point is: she was still very intrigued. She helped me organize and participated in a one-day workshop given by Evan Maletsky at San Juan College and a five-day workshop presented by Heinz-Otto, Evan, Terry, Dietmar, and Dr. Christopher Barton of the USGS. She reads vociferously about fractals and plays on the computer and her graphing calculator. She's been successfully preparing science fair projects on fractals since she was in the seventh grade. At the time of this writing, she is putting the finishing touches on her science fair project to compete, as a Sophomore in high school, at the International Science and Engineering Fair in Tucson, AZ. Kelley is not a genius nor a child prodigy. She has not yet taken calculus. Kelley's a bright young lady who makes good grades, dances, plays percussion in the high school band, drives an old raspberry red pickup, and tries to push her curfew to the limit. She is simply a young teenager who has been "turned on" to the beauty, to the "simplicity, yet complexity" of the mathematics of dynamical systems, and to the indisputable evidence of fractals and their applications in her present and future world. She recognizes that she is on the brink of a mathematical revolution with these emerging concepts about dynamical systems, and she, even with her youth and limited knowledge, feels the excitement and tremendous power of the fractal mathematics that she's been studying.

Whether we are first graders filled with awe about our world; bright, spunky teenagers ready to challenge life; teachers trying to prepare our students for an ever changing future; research mathematicians discovering new mathematical relationships; or scientists finding applications for these new relationships, there is an appropriate beginning and level of complexity at which we can begin our study about fractals, chaos, and dynamical systems. The point is that we must indeed begin. We have been given a tremendous legacy in the form of these materials, of which this book is the third in the series.

We thank each of you, Heinz-Otto, Evan, Terry, Dietmar, and Hartmut, for the research you've done, the discoveries you've made, and the lectures, workshops, and written materials you've prepared so that we all may have a better understanding and appreciation of the world in which we live. We look forward to many more great accomplishments from you in the future. And to you, Lee Yunker, we are grateful to you for your caring, invaluable contributions to our lives and our education, and we miss you.

Jana Wallace
Math Instructor

San Juan College
Farmington, New Mexico

Unit 7
IFS in Two Dimensions

KEY OBJECTIVES, NOTIONS, and CONNECTIONS

Activities contained in this unit discuss repeated mappings in two dimensions. The fact that the Sierpinski Triangle appears as a recurring content theme emphasizes the fact that the same limit figure may emerge from more than one iterative process. Moreover, as suggested by Activity 7.4, although the geometry of the limit figure might be represented in more than one way, these representations may be intimately inter-related.

Connections to the Curriculum

The material covered in these strategic activities integrally relate to content already included in contemporary mathematics programs. Consequently, these activities can be invested into the curriculum as stand alone units, or they may be periodically integrated into existing content sequences as applications or enrichment activities.

PRIMARY CONNECTIONS:

Mappings	Function Composition
Geometric Transformations	Visualization
Geometric Convergence	Number Patterns
Combinations	Iteration and Recursion
Pascal's Triangle	Similarity

SECONDARY CONNECTIONS:

Binary Numbers	Affine Transformations
Limit Concept	Convergene
Binomial Expansion	Addressing
Geometric Patterns	Graphing Calculator

Underlying Notions

Chaos Game

An iterative process that plots successive points within randomly selected subregions of a polygon is called a Chaos Game. The location of a given point of the sequence within a selected subregion depends upon the location of its predecessor within the encompassing polygon.

Address

Addresses identify subregions of an image at a given stage of an iterative construction process. At any finite stage the address consists of a finite string of symbols. Infinite extensions of such strings yield addresses of points in the fractal.

Transformation

When points in one geometric figure are mapped to the points of another figure, the mapping is referred to as a transformation. A transformation is affine if the mapping is a combination of linear mapping and a translation.

MATHEMATICAL BACKGROUND

The Bigger Picture

A number of classical approaches exist for modeling the processes or artifacts present in natural phenomena. These approaches include using static formulas to represent the relationship between variables operative within the phenomena. Alternately, the techniques of the calculus and differential equations can be employed to more dynamically monitor rates of change between two or more variables. Unfortunately, in order to successfully apply these classical models to natural settings, the phenomena must generally display relatively smooth and continuous behavior.

Iterated function systems offer a new and process oriented approach to studying the dynamics of phenomena already ammenable to the classical approaches. However, in addition, iterated function systems also provide an avenue for modeling certain phenomena that fail to satisfy the classical restraints. Disruptive phenomena that fail to be continuous, processes that exhibit turbulence, and mechanisms containing apparently random events all exist as excellent candidates for analysis by iterated function systems.

In this unit we investigate basic notions of mappings, addressing, and iteration involved in iterated function systems.

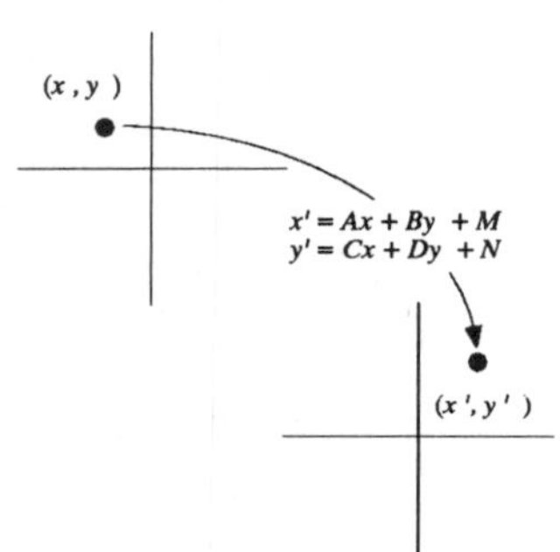

IFS in Two Dimensions

In algebra two equations of the form

$$x' = Ax + By + M$$
$$y' = Cx + Dy + N$$

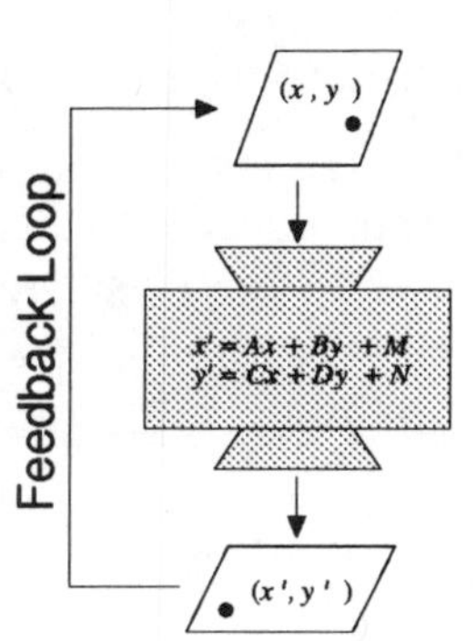

are called a linear system. From a geometric point of view, equations of this form define an *affine* mapping from the set of all points $\{(x, y)\}$ to an image set $\{(x', y')\}$. If, in addition, the value of $AD - BC$ is nonzero, then the mapping is one-to-one, and invertible. In this case, the affine transformation is a 1-1 function that assigns to each point (x, y) a unique image point (x', y') and for each point (x', y') in the plane there is a unique point (x, y), which gets mapped to (x', y').

Now supppose that the transformation is allowed to operate on some initial point (x, y) or figure so as to produce an image point (x', y') or image figure. By repeatedly transferring the last image back to the domain of the mapping, a *feedback loop* is established. Accordingly, each iteration of the feedback process produces a new image.

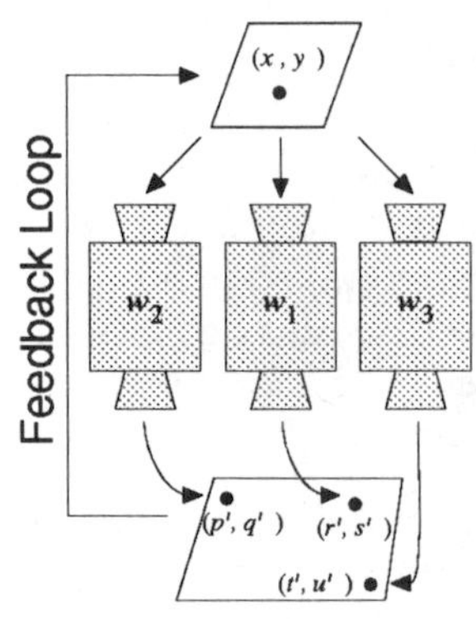

Finally, imagine that each point or image is simultaneously passed through more than one transformation. Suppose that three mappings w_1, w_2, and w_3 are operative in the transformation process. Then the combined effect of the three transformations applied to a set of points is to produce a collage consisting of the three resultants separately generated by each transformation. The transformations could be collectively referred to as a *function system* in the same way that multiple equations are collectively called a system of equations.

The combination of the function system and a feedback process together defines an *Iterated Function System*, commonly denoted by the abbreviation IFS. In short, in an iterated function system more than one transformation repeatedly acts upon points or figures to produce a sequence of image points or image figures.

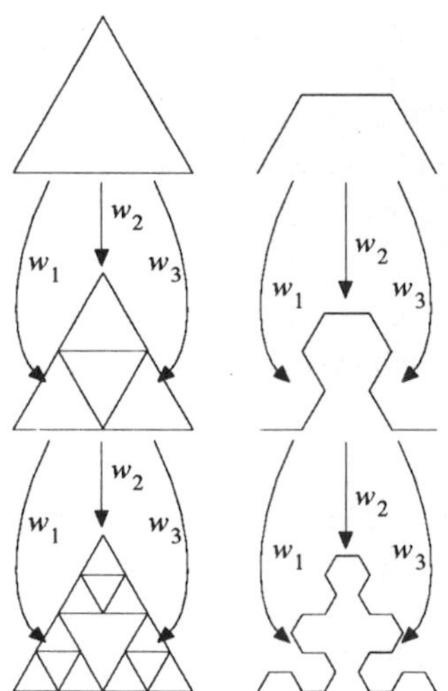

The construction of the Sierpinski Triangle provides an example of such an IFS process. Each of three affine mappings reduces the domain figure by a factor of 1/2, and then maps the figure so as to form one of the three parts of the image figure. The image figure subsequently becomes the new domain figure and the mapping process repeats. As this iteration mechanism continues, the image figure approaches the Sierpinski Triangle as the limiting state.

The construction process invests into the Sierpinski Triangle a very clear *self-similarity* structure. Specifically, the final state image can be partitioned into parts in such a way that each of these parts is a replica of the whole image. This feature apparently derives from the mapping process in which each mapping contributes a copy of the whole image at any stage to a part of the image figure at the next stage. Accordingly, the final state image is self-similar if it can be partitioned into parts in such a way that each of these parts is the image of the whole figure under an affine mapping.

Like the Sierpinski Triangle, the Sierpinski Curve can be generated by an iterated function system. Let mappings w_1, w_2, and w_3 each reduce the domain figure at each stage by a factor of 1/2, and further suppose that the mappings w_2 and w_3 perform a flip at the vertical axis through the middle of the figure and a counterclockwise rotation of 240 degrees (w_2) or 120 degrees (w_3). Then as w_1, w_2, and w_3 repeatedly place copies into the regions assigned to each of them, the sequence of image figures approach the Sierpinski Curve as the limiting state. In fact, applying these same mappings to a triangle at the initial stage or to the figure used to generate the Sierpinski Curve results in a limit state figure that is the same in both cases. The points contained in the Sierpinski Triangle and those contained in the Sierpinski Curve ultimately form the same set.

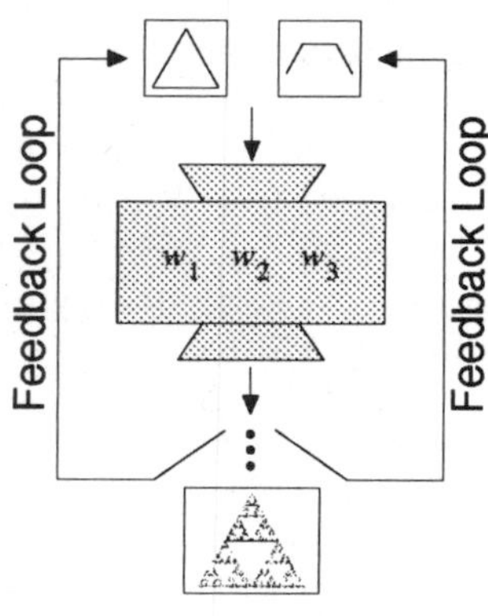

A variety of *addressing* techniques provide mechanisms for identifying points in the final state figure. Addresses not only provide a means for determining whether or not a given point is an element of the final image set, but they also offer a symbolic description of the final figure. Such a characterization can aid in analyzing the cardinality of the set. Once determined, a known characterization can be modified slightly so as to generate new fractals as well as their attendant construction processes. In essence, whole families of fractals can be summarized by features common to their addressing attributes.

In these regards, *Kummer's Criterion* represents one way in which a numerical addressing scheme can provide an elegant approach to understanding a whole family of fractal images. Suppose that the coefficients in Pascal's triangle are recorded as remainders upon division by some prime number p, and that the addresses along the margins of the triangular array are written as base p numerals. In each row of Pascal's triangle, shade all cells containing the remainder 1. As the number of rows increases, the resulting image formed by the shaded cells approaches a final state image that is self-similar and fractal in regards to that property.

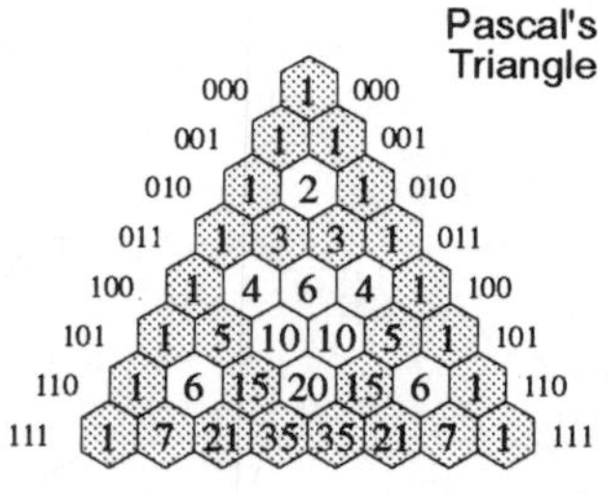

For example, if remainders are recorded upon division by prime number $p = 2$ for a given number 2^n of rows in Pascal's triangle, the shaded remainder 1 cells form an image that corresponds to the stage n Sierpinski triangle.

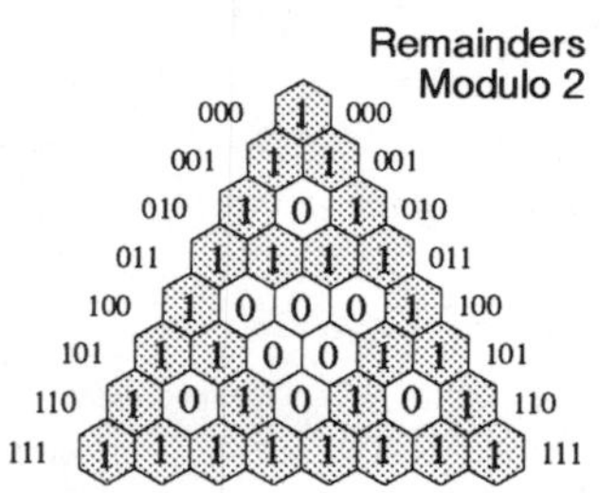

Unfortunately, as the number of rows in Pascal's triangle increases, the magnitudes of the coefficients also increase making their computation cumbersome at best. Kummer's Criterion offers a means for determining whether or not an arbitrarily selected cell will contain a 0 upon division by prime p. Specifically, Kummer proved that prime number p divides $\binom{k+n}{k}$ if a "carry" occurs anywhere in the process of adding the two numbers k base p and n base p that form the cell's address. In short, since Kummer's Criterion is decisive for all prime numbers p, a whole class of fractal images can be characterized by the simple mechanism it provides. For each prime number p, as the number of rows in Pascal's triangle increases, the image formed by the nonzero remainder cells upon division by p (identifiable by Kummer's Criterion) approaches a self-similar final state image that is fractal. Moreover, Kummer's criterion provides the mathematical link between the two characterizations of the Sierpinski triangle given by the addressing scheme on the one hand and the shaded cells in the Pascal triangle on the other hand.

Additional Readings

Chapter 5 of *Fractals for the Classroom, Part One,* and chapter 9 of *Fractals for the Classroom, Part Two,* H.-O. Peitgen, H. Jürgens, D. Saupe, Springer-Verlag, New York, 1992.

USING THE ACTIVITY SHEETS

7.1 The Sierpinski Curve

Specific Directions. Make a copy of TEMPLATE A and TEMPLATE B shown on Activity page 7.1. Cut out the shaded portions of these copies as directed in questions 1 and 4. The actions of moving the cut out pieces onto ILLUSTRATION A and ILLUSTRATION B physically practice the type of mappings that will be used throughout Unit 7. Consequently, this concrete manipulation establishes an experimental basis for activities that follow this first one.

Implicit Discoveries. By repeatedly using three mappings we can construct a curve of increasing complexity. The limiting form of this Sierpinski Curve might be correctly recognized by some as the Sierpinski triangle.

7.2 Why the Chaos Game Works

Specific Directions. The resolution of a TV screen depends upon the number of pixels available for forming an image on the screen. This Activity reviews two equivalent methods for addressing these pixels: Nested Addressing and Composition of Mappings. In order to represent a detailed figure on a screen of a given resolution, each pixel containing a portion of the figure must be turned on. If the addresses of the pixels that should be turned on exhibit a predictable or characteristic pattern, then a chaos game might be employed to randomly find these particular pixels. For example, addresses for a given stage of the Sierpinski triangle construction consist solely of 0's,1's, and 2's. A chaos game generates a random and arbitrarily long string of digits. In order to be effective in replicating the desired figure, this random string must be generated in such a way so as to possess the underlying characteristic implied by the address pattern. In the case of the Sierpinski triangle, this implies generating a random string of 0's,1's, and 2's. Then, as the string length increases the probability approaches 1 that every required address of a given finite length will appear somewhere in the string.

Extensions. Label a 1 on one face of a cube, write a 2 on two faces, and label the remaining three faces with the digit 3. Repeatedly roll the die to obtain a random string of 0's,1's, and 2's in which the probabilities of the respective digits are not equal. Use the ADDRESS COMPUTER to identify pixels on an appropriate pixel grid in order to determine what effect, if any, changing the probabilities has upon the resulting image.

7.3 Addressing the Sierpinski Gasket and Binary Numbers

Specific Directions. This activity practices four distinct approaches to addressing the Sierpinski Gasket. After introducing the NESTED ADDRESSING method, we examine addresses based upon the rectangular coordinate system using both DECIMAL coordinates and BINARY coordinates. Finally, all three of these addressing methods are integrated and shown to be related to addresses that depend upon compositions of MAPPINGS. The development presents an algorithm that allows for a rapid translation from one addressing system to any one of the other three.

Implicit Discoveries. Although a visual representation of the Sierpinski Gasket reveals a striking pattern and a deep self-similarity structure, an elegant symbolic characterization of the Sierpinksi Gasket also presents itself within the formal addresses. Specifically, shaded cells in the Gasket correspond to compositions of mappings that omit w_3, or equivalently, the binary coordinates of shaded cells in the Sierpinski triangle (written vertically above each other) are characterized by never having 1's simultaneously appearing in the same column.

7.4 Combinations

Specific Directions. While this Activity develops three methods for computing the number of combinations that can be formed by selecting r elements from a set containing n elements, its primary objective is to introduce the array of coefficients that form Pascal's triangle from two distinct perspectives. On the one hand, any element in a given row of Pascal's triangle can be computed by adding the two entries immediately above it in the prior row. In this sense, we can compute the entries in Pascal's triangle by an iterative addition process that determines one row after another. On the other hand, a given coefficient can be computed directly by evaluating an appropriate combination. Specifically, denoting the entries of row n from left to right as entry 0, entry 1, entry 2, and so forth, the formula $\binom{n}{r} = \frac{n!}{r!(n-r)!}$ computes the value of entry r in row n. In effect, the row and entry number act as a sort of address for a particular entry within Pascal's triangle.

Implicit Discoveries. In Activity 7.3 the Sierpinski Triangle emerged as the end result of an iterative process that involved finding midpoints and deleting central triangles. Individual points within the final state image were also shown to be identifiable by ternary addresses. In an analogous manner, while the coefficients of Pascal's triangle emerge from an additive iterative process, they can also be individually and directly computed by knowing the location of any particular desired entry within the larger array.

7.5 Sierpinski's and Pascal's Triangle via Combinations

Specific Directions. The taxi problem that opens this activity is a disguised version of the falling marble problem that opened Activity 7.4. In the problems 1 and 2 of Activity 7.4, a marble will bounce from any peg in one of two possible directions. In problem 1 of this activity, a taxi driver will choose at any given corner to proceed in one of two possible directions. Experiments are called binomial when they involve a sequence of n identical choices having exaclty two possible outcomes on each choice. Upon counting the number of ways a particular result could occur in a sequence of such choices (such as the marble falling onto a particular peg, or the taxi visiting a particular corner), the coefficients of Pascal's triangle emerge. The key observation is that shading the cells of Pascal's triangle containing combinations that are not divisible by 2 gives rise to the patterns identical to those obtained in the construction of the Sierpinski triangle.

Since familiarity with a variety of methods for computing the coefficients of Pascal's triangle is an essential prerequisite for this Activity, first complete Activity 7.4. You should also review addition in bases other than base 10.

Extensions. In this Activity Pascal's triangle provides a basis for coloring patterns that give rise to a whole class of fractal images. One can shade in cells with combinations that are not divisible by a number different from 2, e.g., 3. See also Activity 1.10 and 1.11 in Strategic Activities Volume One.

7.6 Kummer's Criterion

This Activity provides the mathematical link between the Sierpinski triangle and Pascal's triangle. Recall the results of Activities 7.3 and 7.5:

• The binary coordinates of shaded cells in the Sierpinski triangle (written vertically above each other) are characterized by never having 1's simultaneously appearing in the same column.

• Sierpinski's triangle can be obtained by shading those numbers from Pascal's triangle that are not divisible by 2.

These two characterizations be seen to be equivalent by Kummer's criterion: the prime number p divides $\binom{k+n}{k}$ if a "carry" occurs anywhere in the process of adding k and n as base p numbers.

In this Activity this fact is explored using some combinations and the figures familiar from the Sierpinski triangle construction.

Moreover, computation of these coefficients of Pascal's triangle quickly becomes unwieldy. Kummer's Criterion supplies a convenient method for determining whether or not a particular entry anywhere in Pascal's triangle should be colored.

Implicit Discoveries. Kummer's criterion is the mathematical basis for the description of the Sierpinski triangle using the shading rule for Pascal's triangle.

7.7 Mappings

Specific Directions. The objective in this activity is to investigate the geometric properties of one-to-one affine transformations. We consider affine mappings that multiply length by a constant ratio r in one direction and by a different constant s in another direction. However, when the ratio r of expansion or reduction is constant for all mapped dimensions, the affine mapping is referred to as a similarity. In this case, the image figure is a proportionally reduced or expanded copy of the original.

7.8 Copying Machines Gone Wild

Specific Directions. In an iterated function system more than one transformation repeatedly acts upon points or figures to produce a sequence of image points or image figures. In this activity all of the mappings are affine transformations. Some of them are also similarities in which the ratio r of expansion or reduction is constant for all mapped dimensions. Others reduce mapped lengths in one direction by a different constant than that applied to lengths in a separate direction.

Implicit Discoveries. A systematic application of an iterated function system to an original figure can generate a self-similar fractal final image that replicates an artifact such as a plant common to the natural world.

7.9 The Sierpinski Triangle, Sierpinski Curve, and Self-Similarity

Implicit Discoveries. The mappings that comprise the iterated function system used to produce the Sierpinski Curve also generate the Sierpinski triangle. In the final state image all copies of the original image converge to points. In this sense, the details of the original figure have no bearing upon the final outcome.

7.1 THE SIERPINSKI CURVE 7.1A

During the early 1900's, the eminent polish mathematician Waclaw Sierpinski did pioneering work involving fundamental properties of curves in space. The following activity offers a path into results that relate to Sierpinski's important work.

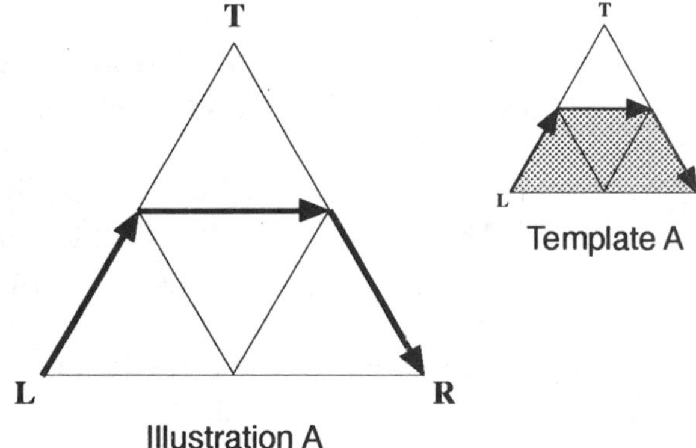

Illustration A

1. Make an exact copy of TEMPLATE A. Cut out the shaded portion from your copy to use as a pattern.

2. Place the TEMPLATE A pattern that you made in problem 1 onto the lower left section of ILLUSTRATION B. Make sure that the TEMPLATE A arrows match those of ILLUSTRATION B. This may require a flip.
   Likewise, move the TEMPLATE A pattern onto the top section and then the lower right section of ILLUSTRATION B. In each case make sure that the TEMPLATE arrows match those of ILLUSTRATION B.

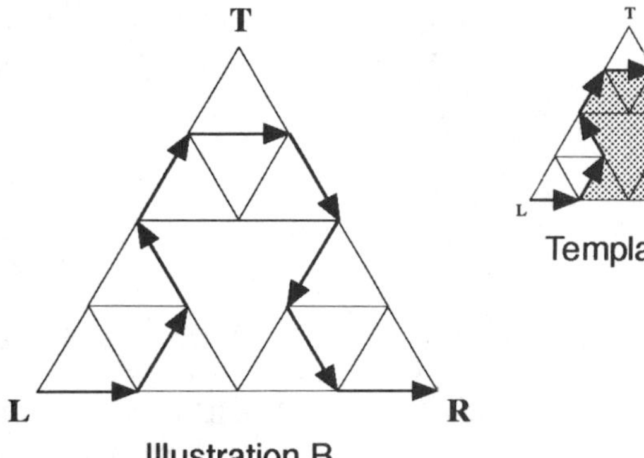

Illustration B

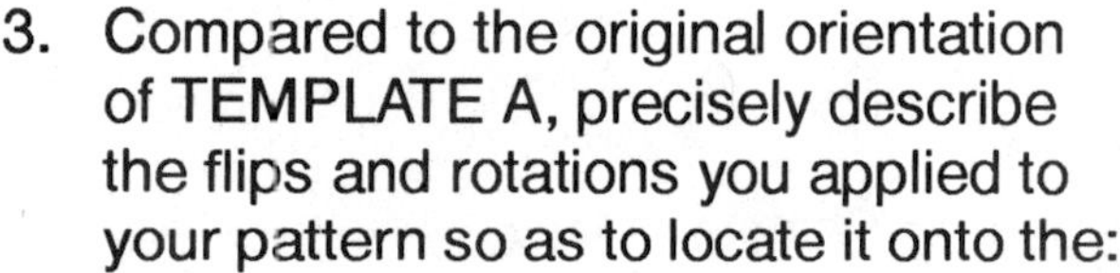

3. Compared to the original orientation of TEMPLATE A, precisely describe the flips and rotations you applied to your pattern so as to locate it onto the:
   a. Lower left section of ILLUSTRATION B.
   b. Top section of ILLUSTRATION B.
   c. Lower right section of ILLUSTRATION B.

4. Make an exact copy of TEMPLATE B. Cut the shaded portion out from your copy to use as a pattern including the arrows at the left and right bottom.
   a. From its original orientation on TEMPLATE B, flip your pattern over using the vertical line through vertex T as an axis. Then perform a 240 degree counterclockwise rotation.
   b. Place the pattern in this new orientation into the lower left section of ILLUSTRATION C in the same way that you placed TEMPLATE A onto ILLUSTRATION B.

7.1B

c. Trace along the edge of your pattern in order to draw a portion of a curve within the triangular skeleton provided by ILLUSTRATION C.

d. Repeat steps a–c for the lower right section of ILLUSTRATION C. However, in this case, after flipping the TEMPLATE B, perform a 120 degree counterclockwise rotation.

e. Based upon your results in problem 3, properly orient TEMPLATE B within the top section of ILLUSTRATION C and trace the pattern.

This should result in a continuous path along the curve from point L to point R.

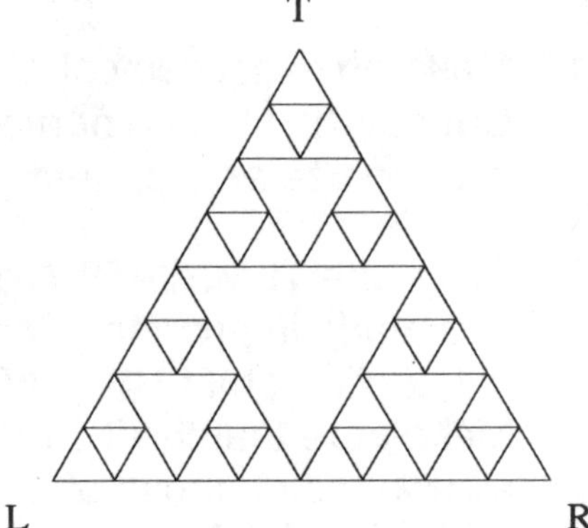

Illustration C

5. As shown by the pictures below, the process that you used to create the curve drawn in ILLUSTRATION C can be continued through many more stages. Write a detailed algorithm that describes how the next stage of the curve can be obtained by making half size copies and applying appropriate flips and rotations.

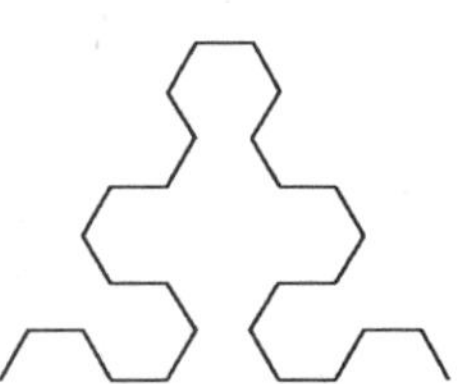
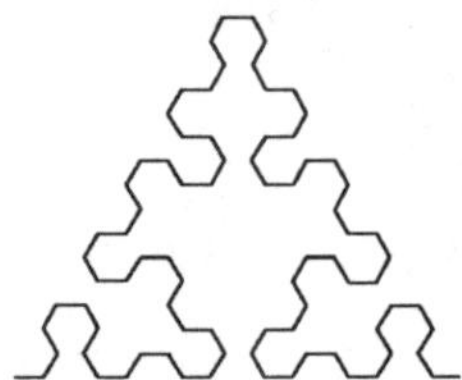
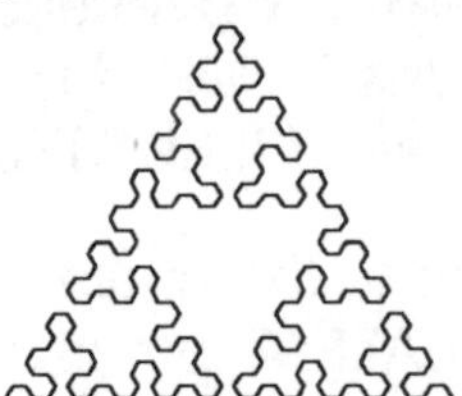

Three equal segments form the curve displayed in ILLUSTRATION A. Suppose that each of these segments has a length of 1 unit thereby giving the curve in ILLUSTRATION A a total length of 3 units. The segments that form the curve in ILLUSTRATION B are one half as long as the segments used in ILLUSTRATION A.

6. After completing the following table:

Stage	1	2	3	4		n
Segment Length	1	1/2				
Curve Length	3					

a. Write a recursive formula that computes the length of the curve at stage n from the length at stage $n - 1$.

b. Write an explicit formula that computes the length of the curve at stage n.

7. As the stage number n approaches infinity, what length is approached by each segment in the curve?

8. As the stage number n approaches infinity, what limit value is approached as the total length of the curve?

7.2 WHY THE CHAOS GAME WORKS

7.2A

The screen of a typical TV has multiple rows of many small phosphor dots called pixels that can be turned on or off by an electrical charge in order to form a picture. On the TV's shown to the right, each small square represents one pixel. A TV having many pixels has a higher resolution than one having fewer pixels. Accordingly, the TV shown on the left has a lower resolution than the one shown to the right.

In order to systematically give names to the pixels, we used a NESTED ADDRESSING approach as illustrated to the right. The small, section 012 black cell shown in the bottom figure is contained within subregion 01 shown in the middle diagram which itself is in section 0 of the whole image. The NESTED ADDRESS of the small black cell is 012.

Notation: $Q_{012} \subset Q_{01} \subset Q_0$

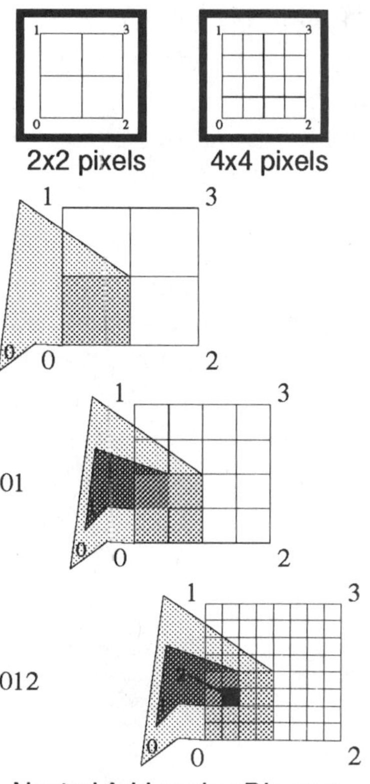

Nested Addressing Diagram

1. In each of the grids shown below, write the NESTED ADDRESS on each blank pixel.

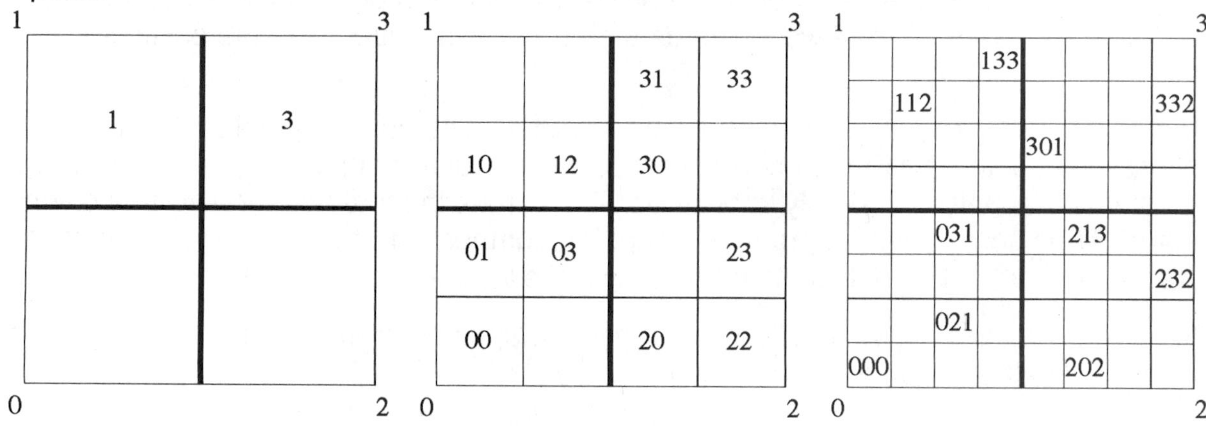

2. Make a "Nested Addressing Diagram" like the one preceding problem 1 to illustrate the following Nested Addresses.

   a. $Q_{310} \subset Q_{31} \subset Q_3$ b. $Q_{231} \subset Q_{23} \subset Q_2$

In addition to the NESTED ADDRESS system practiced in problems 1–2, we can also identify a particular cell in a grid by reference to a composition of MAPPINGS.

7.2B

The diagram to the right shows the result given by the composition $w_0 w_1 w_2(Q)$ of the mappings w_2, w_1, and w_0. Each mapping makes a 1/2 sized copy of the large square and places it back into a subsection of itself.

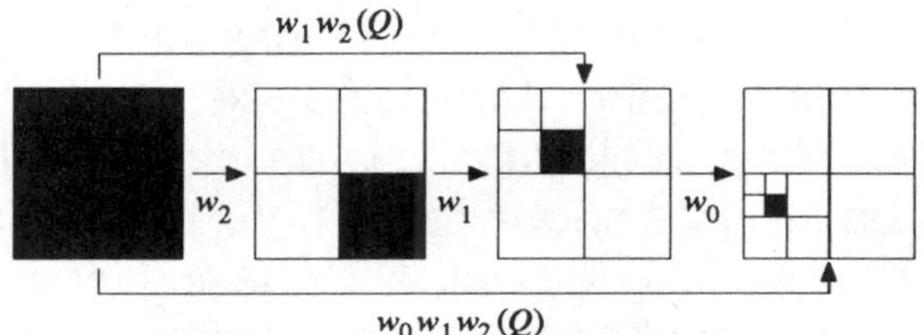

3. As illustrated in the left diagram below, each mapping copies a given original point to a corresponding image point. The location of the image point depends upon the particular behavior of the individual mappings w_0, w_1, w_2, or w_3. In the middle and right diagrams, plot the image points generated by the separate actions of the four mappings. Hint: Draw lines from the original point to the corners of the large square.

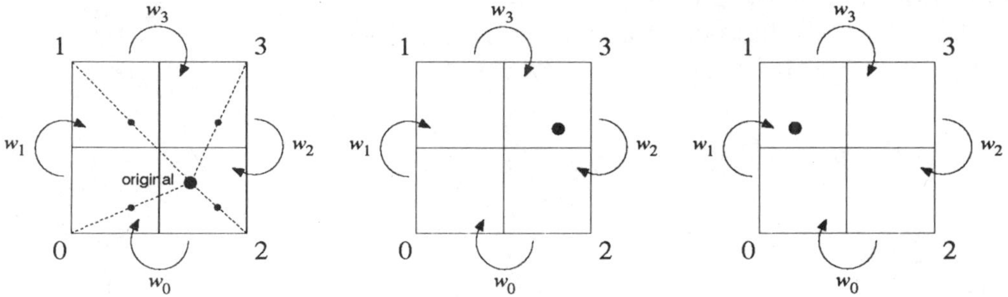

4. Each image point in problem 3 divides a line segment joining the original point to a corner of the large square into two subsegments. What is the ratio between these two subsegments?

In effect, every point in the original square is mapped to a point within a subsquare half as large and precisely half way towards the corner associated with the particular mapping.

In problem 3, by knowing the location of the initial point, you were able to plot subsequent points based upon the action of a particular mapping w_0, w_1, w_2, or w_3. In problems 5–6 below, you only know the specific mapping that is operative at a given stage. Thus, for a given mapping (or mapping composition) there is a unique pixel of a certain resolution that will contain the image of any point under the mapping.

5. Color the two pixels identified by the given mappings or mapping compositions that contains the image point.

a. w_3 and w_2

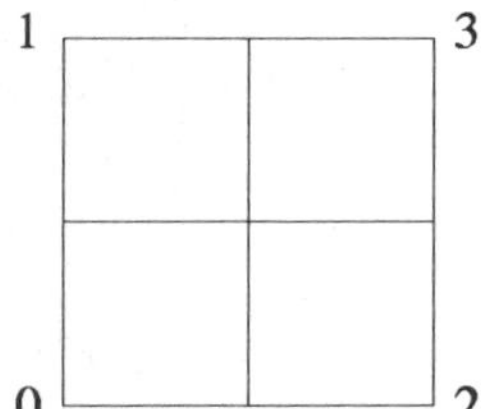

b. $w_3 w_1$ and $w_2 w_3$

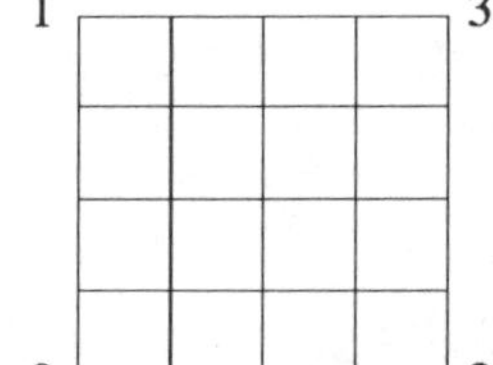

c. $w_3 w_1 w_0$ and $w_2 w_3 w_1$

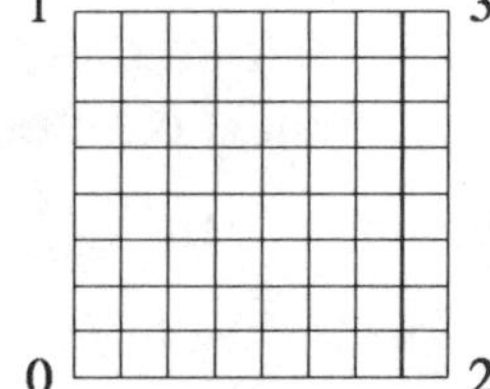

7.2C

6. Compare the results obtained in problem 2a and for $w_3 w_1 w_0$ in problem 5c. In what way does the NESTED ADDRESS of a pixel identify a composition of mappings leading to the same pixel?

Now suppose that an address specifies a very small pixel on a TV, but imagine that the TV has a lower resolution compared to that required by the given address. On a low resolution TV many addresses will lead to the same pixel. For example, as shown to the right, Q_{20}, Q_{21}, Q_{22}, and Q_{23} and all specify the same pixel Q_2.

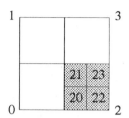

7. Color the pixel identified by each of the following nested addresses or compositions.
   a. Q_{120}, Q_{12}, Q_{122} and $w_1 w_2 w_0, w_1 w_2, w_1 w_2 w_2$
   b. $Q_{1020}, Q_{120}, Q_{1022}$ and $w_1 w_0 w_2 w_0, w_1 w_0 w_2, w_1 w_0 w_2 w_2$

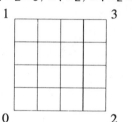

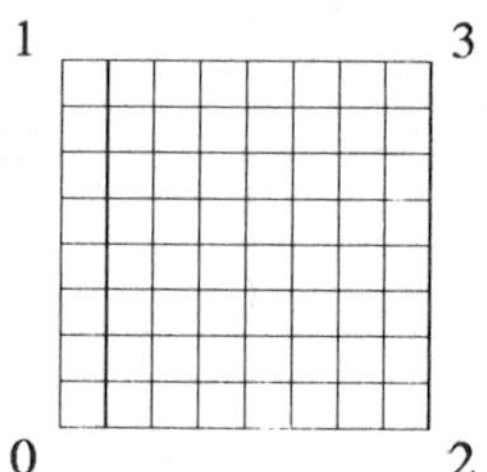

8. If the TV screen has resolution $2^n \times 2^n$ pixels, how many digits in an address are significant for locating a particular pixel?

On the right, we introduce a coloring pattern within a sequence of finer and finer grids. If continued ad infinitum, the coloring pattern leads to the self-similar Sierpinski Triangle.

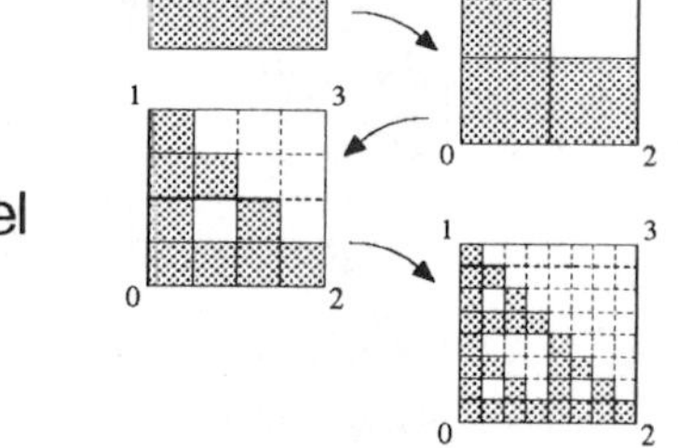

9. How many digits would be significant for addressing a pixel on the next (fifth) screen in the coloring pattern shown to the right?

10. Carefully copy the ADDRESS COMPUTERS shown below onto heavy paper or onto a card. Cut out your copies. Now cut out the L-shaped center sections so as to make an L-shaped window in the middle of each card.

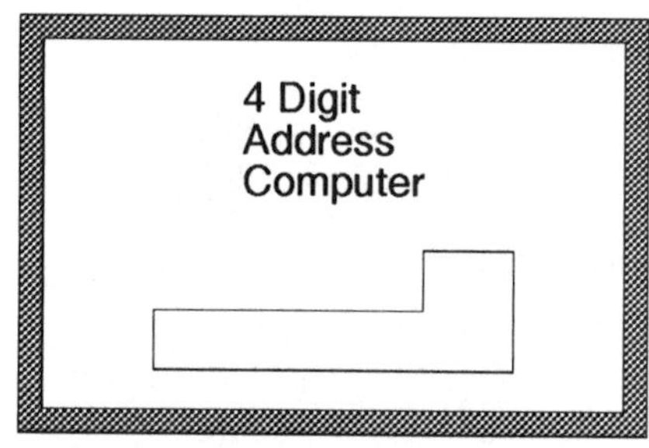

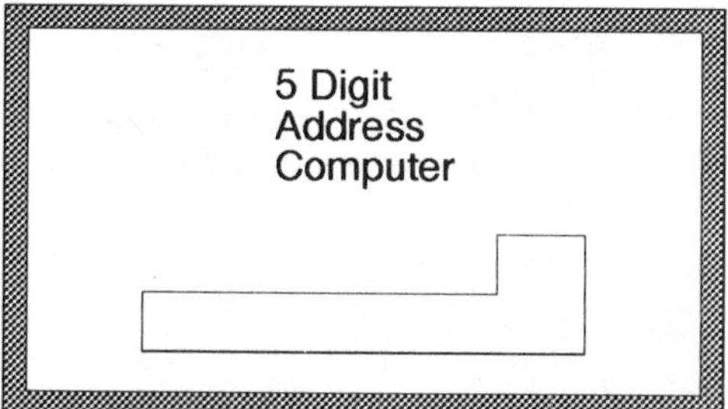

7.2D

11. Place the 4 DIGIT ADDRESS COMPUTER over the right end of the following chart to reveal the first address. Color the indicated pixel in the 16 × 16 pixel grid below chart. Slide the 4 DIGIT ADDRESS COMPUTER one digit to the left to reveal a second address. Color the indicated pixel. Repeatedly slide the ADDRESS COMPUTER window one digit to the left and each time color the indicated pixel.

20	19	18	17	16	15	14	13	12	11	10	9	8	7	6	5	4	3	2	1
2	0	1	2	1	0	0	0	2	0	2	2	1	1	2	0	0	2	1	2

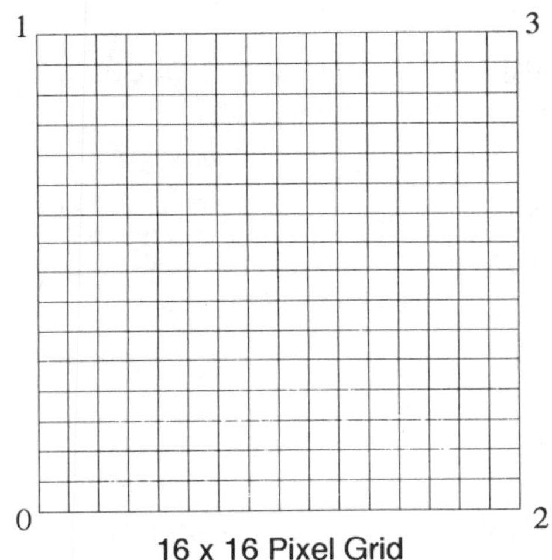

16 x 16 Pixel Grid

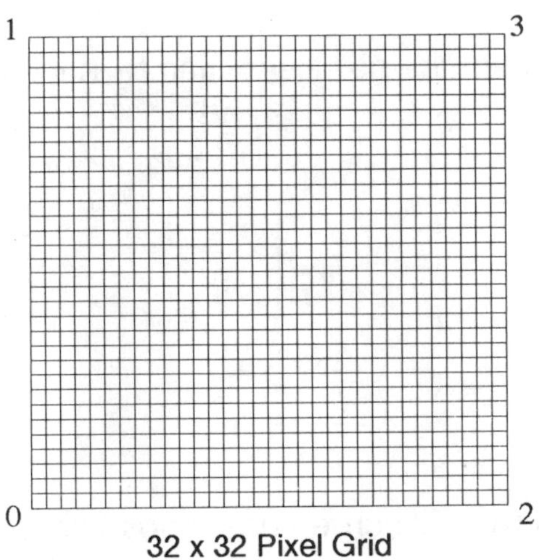

32 x 32 Pixel Grid

12. Mark the faces of a fair die so that two faces display 0's, two faces display 1's, and two faces display 2's. Roll the die 20 times and record the resulting sequence of 0's, 1's, and 2's in the following table.

20	19	18	17	16	15	14	13	12	11	10	9	8	7	6	5	4	3	2	1

13. Follow the instructions in problem 11, but this time use the chart you constructed in problem 12 together with the 5 DIGIT ADDRESS COMPUTER. Color the pixels in the 32 × 32 grid above.

14. Study the coloring pattern illustrated prior to problem 9. In the construction of the Sierpinski Triangle, how many pixels are colored on a TV with resolution:
a. 2×2 b. 4×4 c. 8×8 d. $2^n \times 2^n$

15. Explain why continuing the process practiced in problems 11 and 13 will eventually color all of the required pixels for the stage n Sierpinski Triangle.

16. Summarize how repeatedly mapping a point half way to corners 0, 1, or 2 of a square corresponds to coloring pixels on a TV screen by a list of random addresses of a given length.

7.3 ADDRESSING THE SIERPINSKI GASKET AND BINARY NUMBERS 7.3A

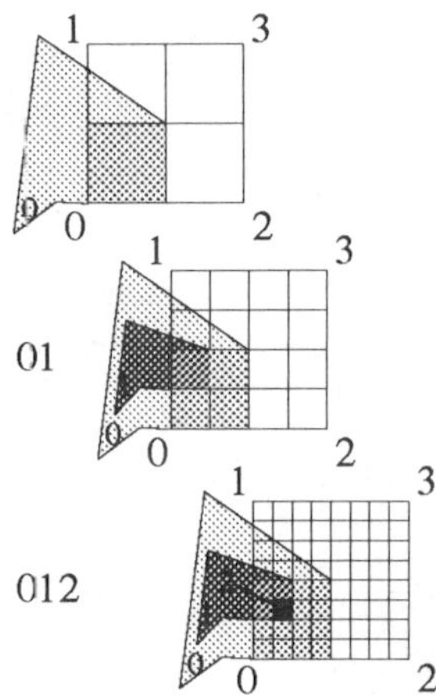

Objects that are called fractal typically possess a property referred to as self-similarity. When an object is self-similar, it can be partitioned into parts that together comprise the whole object in such a way that each of these parts is a replica of the whole object.

In order to give names to all of the parts in a systematic manner, we utilized the NESTED ADDRESSING approach as illustrated to the right. The small, section 2 black cell shown in the bottom figure is contained within subregion 1 shown in the middle diagram which itself is in section 0 of the whole image. The NESTED ADDRESS of the small black cell is 012.

1. In each of the grids shown to the right, write the correct NESTED ADDRESS within each empty cell.

0	2

11	13		33
		30	32
01		21	23
00	02	20	

2. In each grid shown below a single cell is highlighted. What is the nested address for the highlighted pixel?

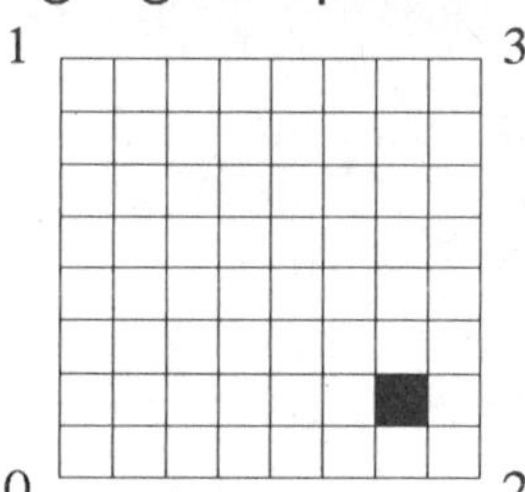

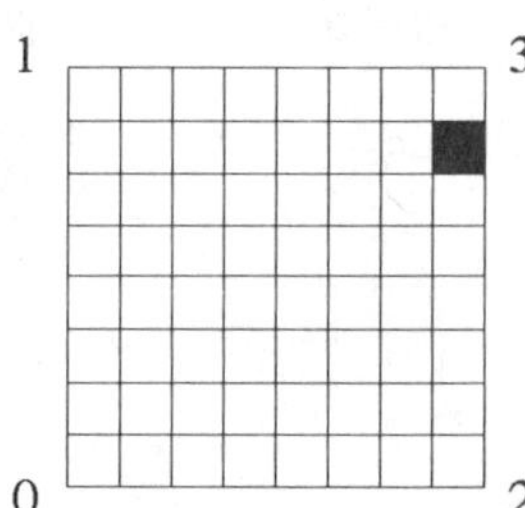

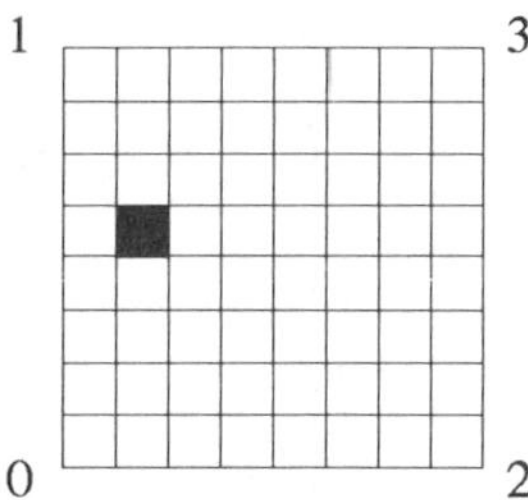

Although a known addressing system may provide a measure of familiarity and simplicity, sometimes a different notation or addressing scheme can offer advantages not present in a prior method.

A common alternative system to the nesting approach reviewed above involves using the rectangular coordinate system. In this system we identify each pixel by a horizontal coordinate x and a vertical coordinate y written as an ordered pair (x, y). The coordinates x and y are expressed in base 10 or in base 2 in the following two exercises.

7.3B

3. In each empty cell below, write the decimal ordered pair that describes the cell's coordinates within the grid.

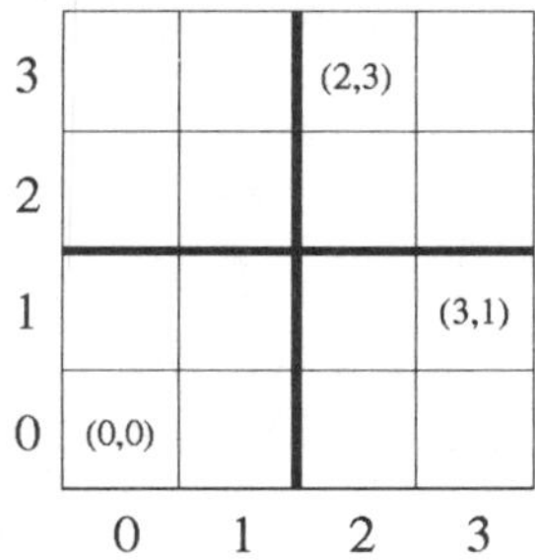

4. In each empty cell below, write the binary ordered pair that describes the cell's coordinates within the grid.

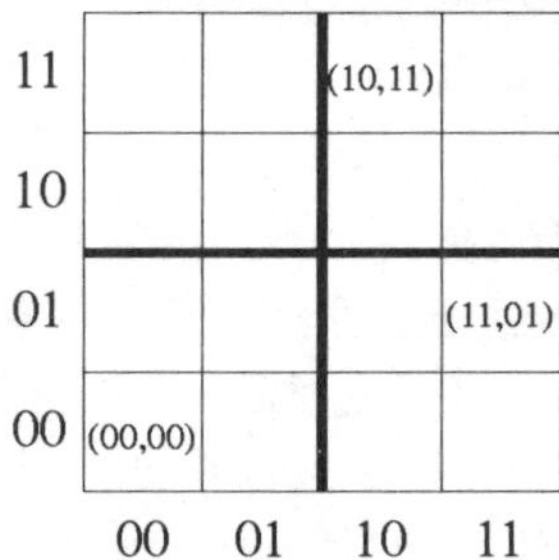

In addition to the NESTED addressing system practiced in problems 1–2 and the COORDINATE addressing system rehearsed in problems 3–4, we also developed an approach in Activity 7.2 for identifying a particular cell in a grid by means of a composition of MAPPINGS.

The diagram to the right shows the result given by the composition $w_0 w_1 w_2(Q)$ of the mappings w_2, w_1, and w_0.

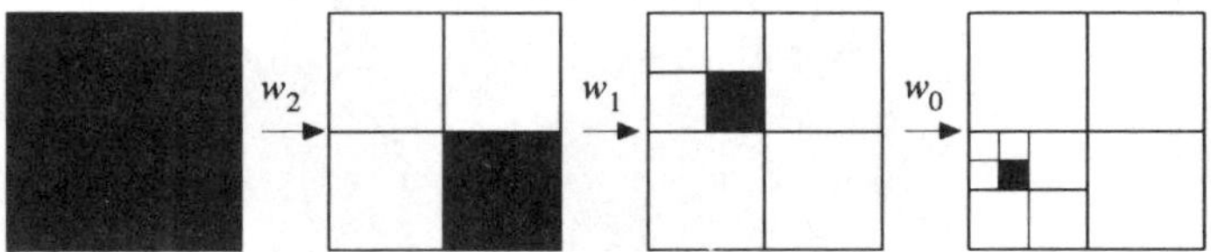

Each of the following screens identifies a single pixel. For each of the highlighted locations the mapping representation of that pixel is already listed. Write the nested address as well as the decimal coordinate and binary coordinate addresses.

5. MAPPING $w_1 w_0 w_2(Q)$

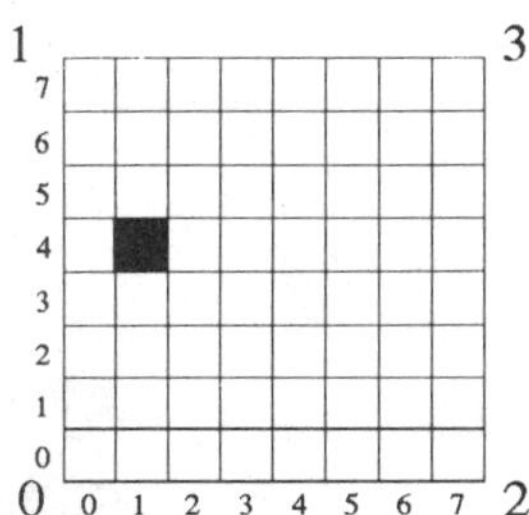

NESTED ________

DECIMAL COORD. ________

BINARY COORD. ________

6. MAPPING $w_2 w_1 w_3(Q)$

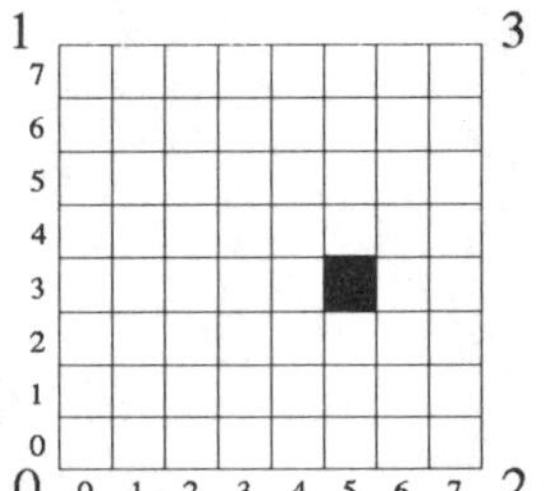

NESTED ________

DECIMAL COORD. (5,)

BINARY COORD. (101,011)

7. MAPPING $w_0 w_1 w_2(Q)$

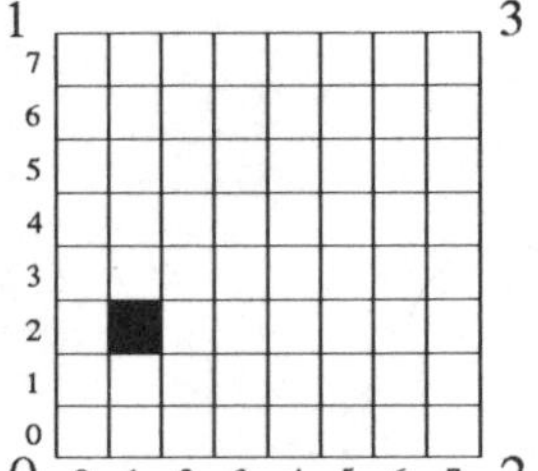

NESTED ________

DECIMAL COORD. ________

BINARY COORD. ________

7.3C

Although a particular notation and addressing system may provide greater ease for some particular application, a striking relationship between all of these approaches allows for a rapid translation from one system to another. The translation steps are supplied in the following algorithm and example.

i. Write the decimal coordinate address.

(5,3)

ii. Convert each decimal coordinate into its binary equivalent.

$5 = 101_{\text{base 2}}$ $3 = 011_{\text{base 3}}$

iii. Form new binary numbers by joining the corresponding digits in the x and y coordinates of the binary address.

$10_{\text{base 2}}$ $01_{\text{base 2}}$ $11_{\text{base 2}}$

Decimal (5 , 3)

Binary (1 0 1 , 0 1 1)

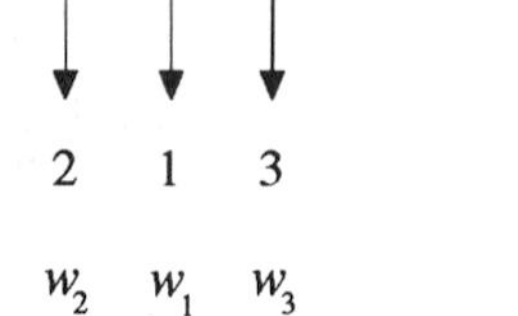

Nested 2 1 3

Mapping w_2 w_1 w_3

iv. Convert each binary number into its decimal equivalent.

$10_{\text{base 2}} = 2_{\text{base 10}}$ $01_{\text{base 2}} = 1_{\text{base 10}}$ $11_{\text{base 2}} = 3_{\text{base 10}}$

8. Use the translation algorithm enumerated above in order to complete the following table.

	a.	b.	c.
Decimal Coordinates	(1,4)		
Binary Coordinates			(001,010)
Mapping Composition		$w_3 w_2 w_1$	

On the next page we will show a coloring pattern within a sequence of finer and finer grids. If continued ad infinitum, the coloring pattern leads to the self-similar Sierpinski Triangle. This deep property of self-similarity with the foregoing addressing schemes will now permit an elegant characterization of the Sierpinski Triangle's structure.

7.3D

In the diagram below, the sequence of three grids establishes a pattern by removing the upper right square from each SHADED box.

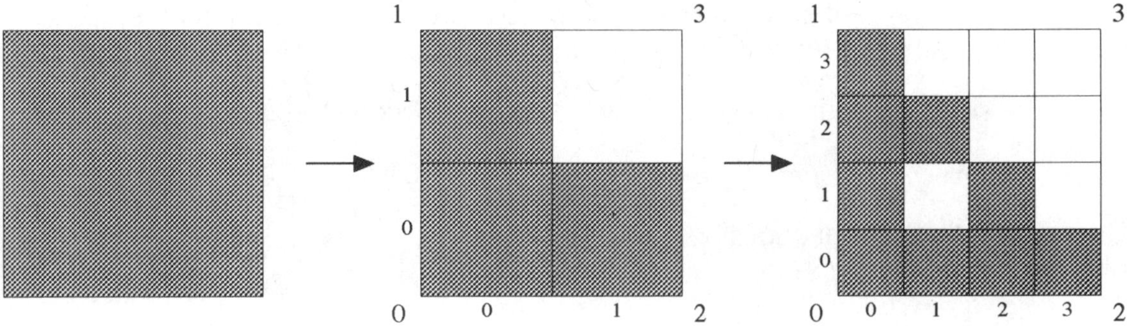

9. Begin with the coarse grids at the top. In each cell of the second and third grid record the NESTED address and the corresponding BINARY coordinate address.

10. What digit 0, 1, 2, or 3, fails to appear in the NESTED addresses associated with shaded cells?

BINARY addresses consist of sequences of 0's and 1's and they all have the form $(a_m a_{m-1} a_{m-2} \ldots a_1 a_0, b_m b_{m-1} b_{m-2} \ldots b_1 b_0)$.

11. For each of the nine shaded cells in the third grid above, write the decimal coordinates and the first binary coordinate above the second binary coordinate as shown below. Record the coordinates for the seven non-shaded cells in the second table below.

SHADED CELLS								
Dec. Coords.	(0,0)	(0,1)						(3,0)
Binary Coords.	00 00	00 01						11 00

NON-SHADED CELLS						
Dec. Coords.	(1,1)	(1,3)				(3,3)
Binary Coords.	01 01	01 11				11 11

12. When binary x- and y-coordinates are written above each other as in problem 11, some binary address pairs never have 1's appearing immediately above and below each other. In other words, in such pairs a_i and b_i are not simultaneously equal to 1. How can the shaded cells in the Sierpinski Triangle be characterized by reference to this binary property?

Result: At any stage of the Sierpinski construction, a specific addressing feature characterizes the self-similarity property of the shaded cells forming the image. Shaded cells correspond to compositions of mappings that omit w_3. Equivalently, *the binary coordinates of a shaded cell never have 1's simultaneously appearing in the same column.*

7.4 COMBINATIONS

7.4A

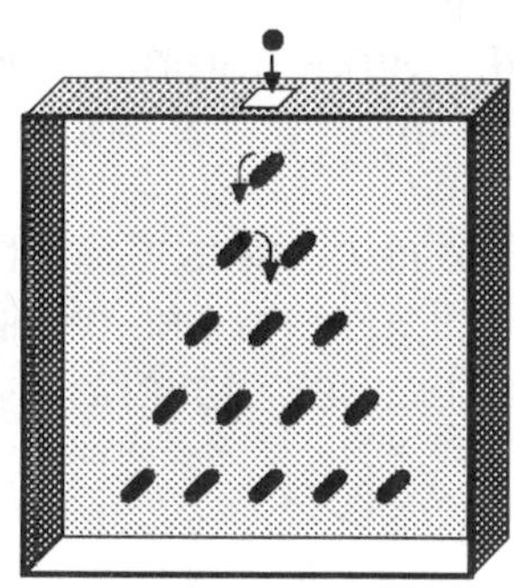

Imagine that we have constructed a box with horizontal pegs configured as shown in the pictures to the right. To discuss the path that a marble will take when dropped through the hole in the top let us introduce a simple mathematical model for this process. When a marble enters the box through the hole in the top, it bounces off of the top peg with an equal probability of falling onto the next lower peg to the left or the next lower peg to the right.

Furthermore we assume the same situation at each peg, i.e., when a peg is hit by a marble, the likelihood that the marble will bounce to the left equals the likelihood that the marble will bounce to the right.

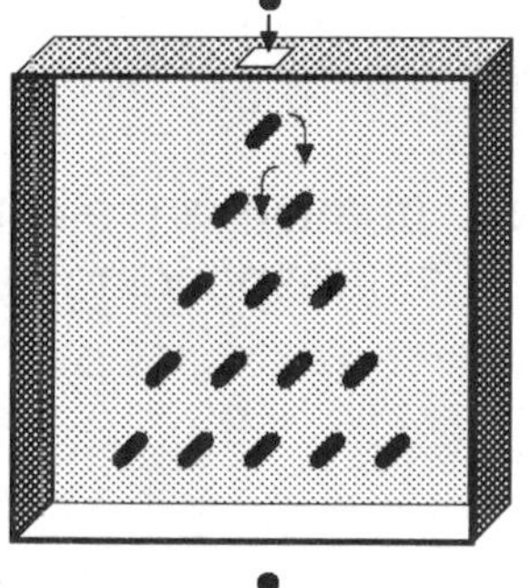

The two boxes shown to the right illustrate the fact that a marble can follow two distinct paths to the middle peg in the third row.

1. The circles in the third box shown on the right correspond to the pegs described above. In each circle write the number of distinct paths that a marble could follow to that peg.

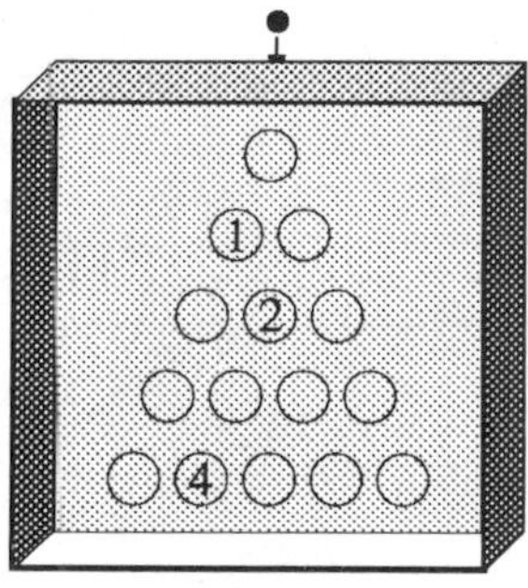

2. If the box was enlarged so as to accommodate another row of pegs, write an addition rule that could be used to compute the number of paths to a particular peg in the bottom row without actually drawing all of the possibilities?

Definition: A *combination* is the number of distinct subsets containing r elements that can be selected from a set S of n elements.

For example, from a set $S = \{x, y, z\}$ of three letters, we can form three distinct two element subsets: $\{x, y\}, \{x, z\}, \{y, z\}$

3. Let set $S = \{w, x, y, z\}$ be a four element set. List all of the distinct subsets containing:
   a. 1 element b. 2 elements c. 3 elements d. 4 elements

In order to express the number of distinct subsets containing r elements that can be taken from a set S containing n elements, we use the notation $\binom{n}{r}$. Since there are 3 combinations of 3 objects taken 2 at a time, we could write $\binom{3}{2} = 3$.

7.4B

4. What number is identified by each of the following expressions?
   a. $\binom{4}{1}$ b. $\binom{4}{2}$ c. $\binom{4}{3}$ d. $\binom{4}{4}$

For large sets S, identifying the number of subsets of a given size becomes a very tedious process of listing all possibilities. However, the following formula allows for a direct calculation. If we define $n!$ to be the product $n \cdot (n-1) \cdot (n-2) \cdots 3 \cdot 2 \cdot 1$, then

$$\binom{n}{r} = \frac{n!}{r! \cdot (n-r)!}.$$

Example: $5! = 5 \cdot 4 \cdot 3 \cdot 2 \cdot 1 = 120$, $\binom{5}{3} = \frac{5!}{3! \cdot 2!} = 10$.

5. In each place in the right hand array below, write the number defined by the corresponding combination in the left hand array. Assume that $0! = 1$.

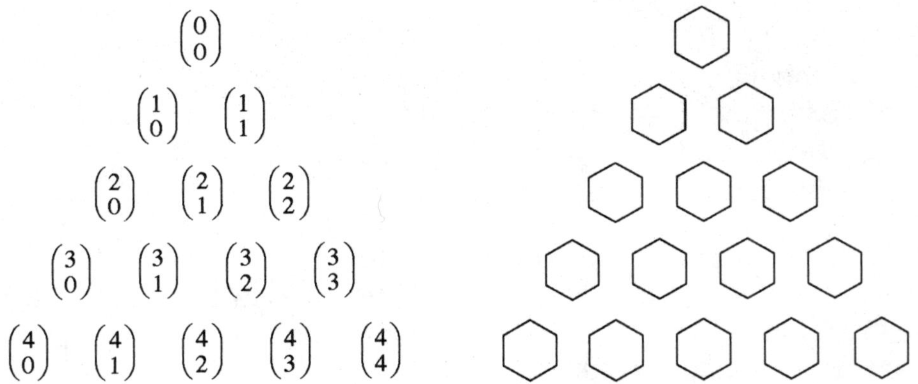

The above triangular array is referred to as Pascal's triangle. Each cell in a row of Pascal's triangle can be computed by adding the two cells above it. However, for combinations involving large numbers, it becomes impractical to extend the grid by sufficiently many rows in order to compute the required combination. Instead, the formula $\binom{n}{r}$ facilitates direct computation of the value in any particular cell.

5. Compute $\binom{8}{2}$ three ways.
   a. List subsets.
   b. Extend Pascal's triangle.
   c. Compute $\binom{8}{2}$ by the formula.

In the next two Activities, combinations as described above and the addressing schemes summarized in Activity 7.3 merge in Kummer's Criterion to provide a powerful tool for creating fractal images of the sort represented by Sierpinski's Triangle.

7.5 SIERPINSKI'S AND PASCAL'S TRIANGLE VIA COMBINATIONS 7.5A

A taxi company serves one quadrant of a city whose streets form a rectangular pattern as shown on the map to the right. From company headquarters at corner A, the dispatcher always sends a cab by the shortest possible route to any given location B. In the example shown to the right, a cab would need to travel a minimum of four blocks from corner A to corner B, though the driver would have a choice of six routes having this minimum four block length. When traced on the map, each of these six routes would take the driver two blocks to the right and two blocks up.

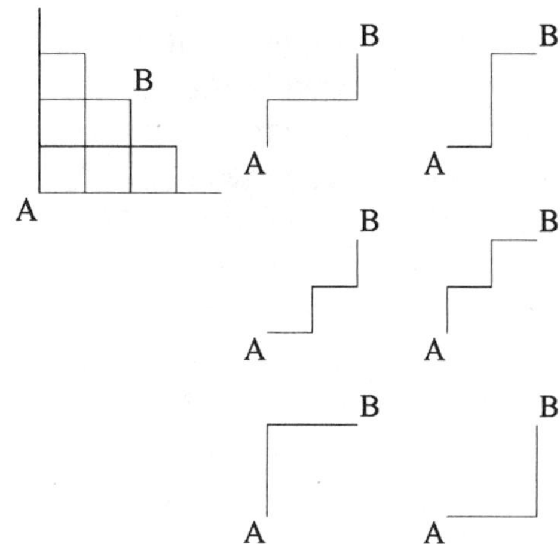

1. Each map below identifies a corner B where a customer is waiting for pickup. How many distinct routes of minimum length are available for a driver to follow from headquarters to each customer?

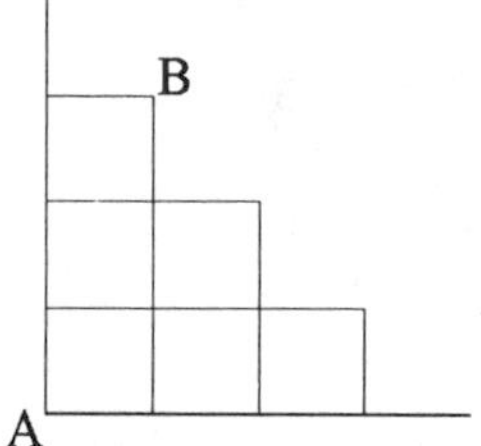

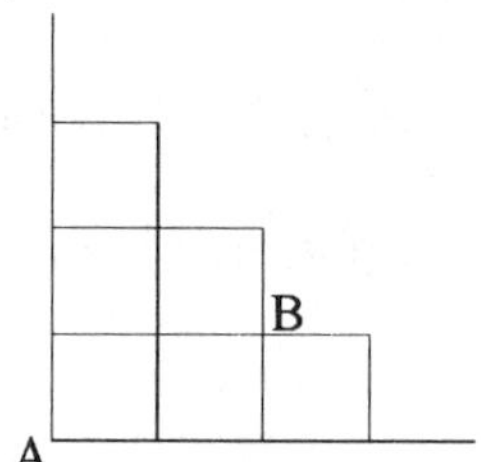

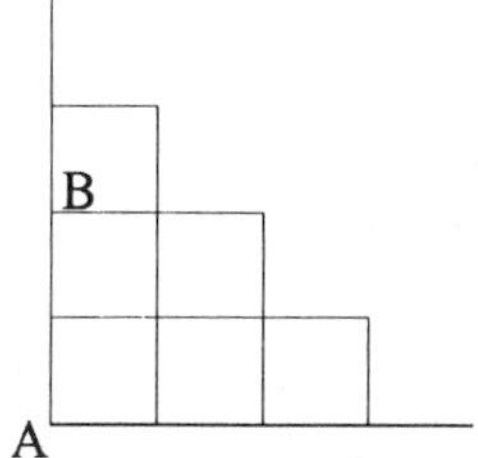

Let n and k denote the number of blocks in a route of minimum length that move the driver to the right and up, respectively. The complete trip consists of $n + k$ blocks. The driver must select k blocks from the total $n + k$ over which the cab will move to the right. The number of ways of making this selection can be computed by the combination $\binom{n+k}{k} = \frac{(n+k)!}{n! \cdot k!}$ or equivalently by $\binom{n+k}{n} = \binom{n+k}{k}$.

2. At each corner marked by a dot, write the combination that represents the number of minimum routes to that corner.

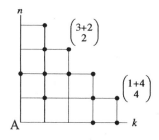

n

$\binom{3+2}{2}$

$\binom{1+4}{4}$

A k

3. Evaluate the combinations that you wrote in response to problem 2. Record the values you obtain on the map below.

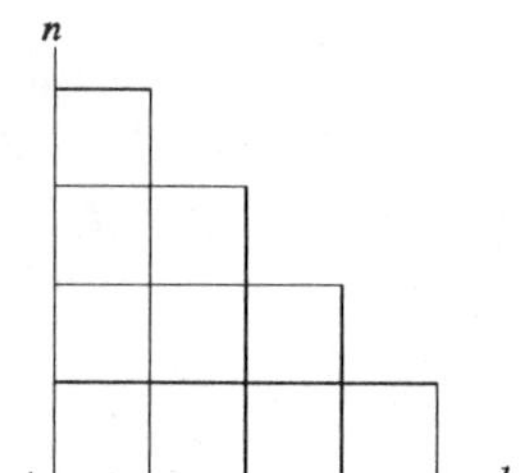

4. You can find the number of paths to a given corner B by adding the numbers associated with two adjacent corners. Where are these adjacent corners relative to the location of B?

7.5B

The figures to the right display the array of combinations referred to as *Pascal's Triangle*. The left array shows Pascal's Triangle in its common orientation. The right figure exhibits the array rotated so as to correspond to the results you obtained in problem 3.

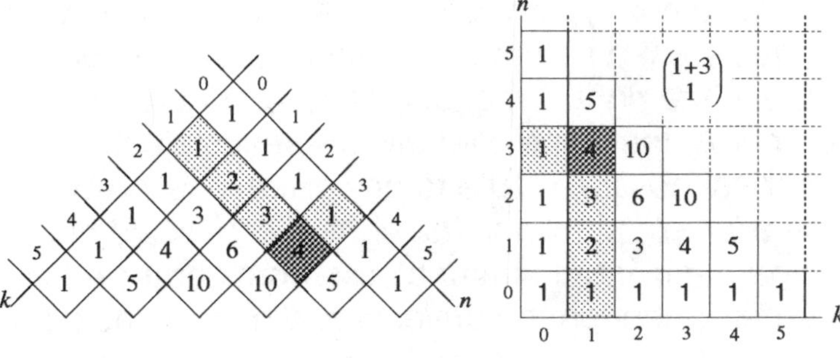

In the previous exercises we have seen how to obtain the number of combinations in these figures. In summary the numbers in the right hand figure can be obtained by the following rules:

Either use the fact that the entry in row n and column k is given by the combination $\binom{n+k}{k}$, or carry out the following algorithm.

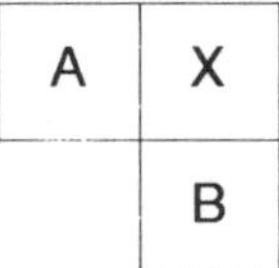

i. The entries at the horizontal k-axis and the vertical n-axis are all equal to 1.
ii. The number contained in cell X is obtained by the sum of the entries of the two bordering cells A and B.

5. Assume that cell X in the above figure is in row n and column k.
   a. What are the combinations associated with the contents of cells A, B, and X?
   b. Write the rule from step ii. of the above algorithm as a formula using the combinations of cells A, B, and X.

6. Verify the formula found in exercise 5b algebraically using the definition $\binom{n+k}{k} = \frac{(n+k)!}{n! \cdot k!}$. Hint: Find the common denominator for the three fractions.

Although the numbers forming Pascal's Triangle appear in many different settings, many people recognize them as coefficients of terms found in binomials raised to various powers.

7. As the powers increase for the following binomials, the multiplication process yields two polynomials that are added together in order to obtain the final expansion. Explain how to obtain these two polynomials.

7.5C

$$
\begin{aligned}
(x+y)^0 &= 1\\
(x+y)^1 &= x+y\\
(x+y)^2 &= (x+y)(x+y) = \quad x^2 + xy\\
&\qquad + \quad xy + y^2\\
&\qquad = x^2 + 2xy + y^2\\
(x+y)^3 &= (x+y)(x^2+2xy+y^2) = \quad x^3 + 2x^2y + xy^2\\
&\qquad + \quad x^2y + 2xy^2 + y^3\\
&\qquad = x^3 + 3x^2y + 3xy^2 + y^3\\
(x+y)^4 &= (x+y)(x^3+3x^2y+3xy^2+y^3) = \quad x^4 + 3x^3y + 3x^2y^2 + xy^3\\
&\qquad + \quad x^3y + 3x^2y^2 + 3xy^3 + y^4\\
&\qquad = x^4 + 4x^3y + 6x^2y^2 + 4xy^3 + y^4
\end{aligned}
$$

In general, if $(x+y)^m = a_m x^m + a_{m-1}x^{m-1}y + \cdots + a_2x^2y^{m-2} + a_1xy^{m-1} + a_0y^m$ then

$$
\begin{aligned}
(x+y)^{m+1} &= (x+y)(x+y)^m\\
&= (x+y)(a_mx^m + a_{m-1}x^{m-1}y + \cdots + a_2x^2y^{m-2} + a_1xy^{m-1} + a_0y^m)\\
&= \qquad a_mx^{m+1} + a_{m-1}x^my + \cdots + a_1x^2y^{m-1} + a_0xy^m\\
&\quad + \qquad a_mx^my + \cdots + a_2x^2y^{m-1} + a_1xy^m + a_0y^{m+1}\\
&\quad b_{m+1}x^{m+1} + b_mx^my + \cdots + b_2x^2y^{m-1} + b_1xy^m + b_0y^{m+1}
\end{aligned}
$$

8. Let $a_kx^ky^n$ be a particular term within the expansion $(x+y)^m = a_mx^m + a_{m-1}x^{m-1}y + \cdots + a_1xy^{m-1} + a_0y^m$. Based upon the values of k and $n = m-k$ which appear as exponents on x and y, what combination should be computed in order to determine the value of a_k?

The polynomial $(x+y)^m$ can be expressed as a sum of terms $a_kx^ky^n$ with $n = m-k$ and coefficients $a_k = \binom{m}{k} = \binom{n+k}{k}$. These numbers are precisely the entries in Pascal's triangle. They form the rows of the left version at the top of the previous page 7.5B and diagonal entries of the right version.

9. If $(x+y)^{m+1} = b_{m+1}x^{m+1} + \cdots + b_2x^2y^{m-1} + b_1xy^m + b_0y^{m+1}$, what coefficients a_k from $(x+y)^m = a_mx^m + a_{m-1}x^{m-1}y + \cdots + a_1xy^{m-1} + a_0y^m$ should be added in order to obtain:
   a. b_2 b. b_3 c. b_i

10. Express the sums in problem 6 in terms of combinations. Explain how this relates to problem 4.

Using the fact that $\binom{n+k}{k} = \binom{n+k}{n}$ we can summarize the results by a formula:

$$(x+y)^m = \binom{m}{0}x^m + \binom{m}{1}x^{m-1}y + \binom{m}{2}x^{m-2}y^2 + \cdots + \binom{m}{m-1}xy^{m-1} + \binom{m}{m}y^m.$$

7.5D

On this page we use the right-angled version of the Pascal triangle as shown. There is a close relationship between the Pascal triangle and Sierpinski's triangle which will be uncovered here.

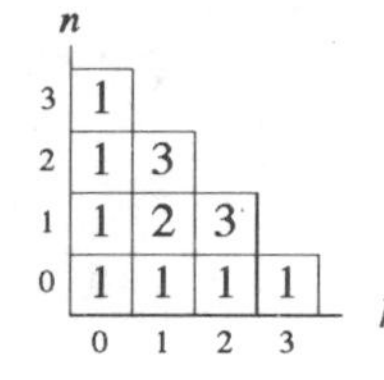

11. For the first four rows and columns shown on the right shade all cells which contain a combination which is not divisible by 2, i.e., shade all cells with odd entries.

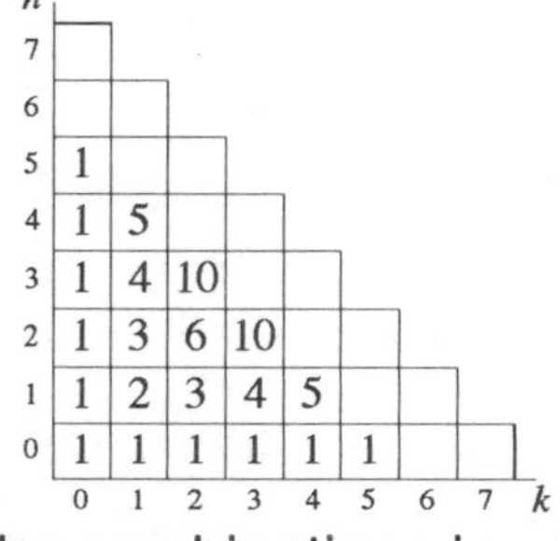

12. Complete the figure on the right writing in all combinations for the first eight rows and columns. Then shade all cells with combinations that are not divisible by 2.

The coloring can be carried out without the explicit knowledge of the combinations in the cells. To see this recall the rule for the algorithmic computation of an entry at cell X: The number is the sum of the neighboring cell contents to the left (cell A) and below (cell B). Thus, the result is not divisible by 2, if and only if one of the cells A or B contains an odd number (i.e., shaded) while the other contains an even number (i.e., not shaded). Therefore, we can color the cells using the following algorithm:

A	X
	B

i. The cells at the horizontal k-axis and the vertical n-axis are all shaded.

ii. The cell X is shaded precisely if A is shaded and B is not or if B is shaded and A is not.

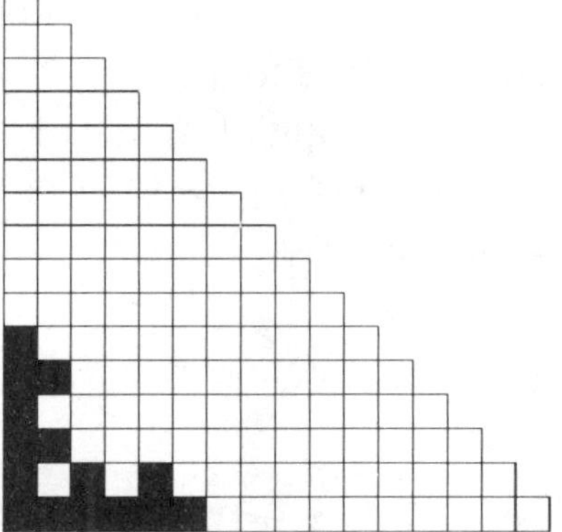

13. Complete the shading of the figure on the right without computing combinations using the above method.

14. The shaded figures obtained in exercises 11–13 are just stages of the Sierpinski triangle generation. What stages do these figures correspond to? How many rows and columns of the Pascal triangle need to be shaded in order to arrive at the i-th stage of the Sierpinski triangle?

This activity has shown that the Sierpinski triangle can be obtained by shading those numbers from Pascal's triangle that are not divisible by 2, i.e., which are odd. The question remains why this is so. The answer is given in the following activity.

7.6 KUMMER'S CRITERION

7.6A

This Activity provides the mathematical link between the Sierpinski triangle and Pascal's triangle. Recall the results of Activities 7.3 and 7.5:

- The binary coordinates of shaded cells in the Sierpinski triangle (written vertically above each other) are characterized by never having 1's simultaneously appearing in the same column.
- Sierpinski's triangle can be obtained by shading those numbers from Pascal's triangle that are not divisible by 2.

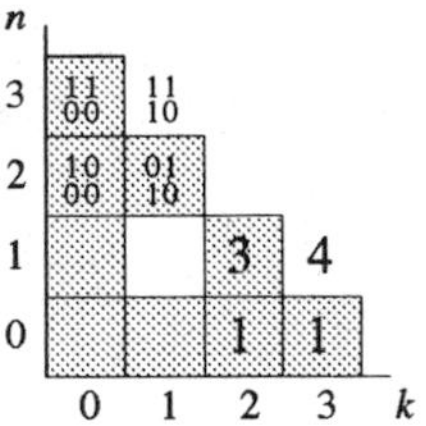

How can these two characterizations be seen to be equivalent?

In 1852 at the University of Berlin, Ernst Eduard Kummer wrote an article that presented a criterion for a given combination to be divisible by a prime number. As suggested by the taxi problem and by the grid to the above, each combination can be identified with a horizontal coordinate k and a vertical coordinate n. Kummer showed that the prime number p divides $\binom{k+n}{k}$ if a "carry" occurs anywhere in the process of adding k and n as base p numbers.

Thus, 2 divides $\binom{k+n}{k}$, precisely if a "carry" occurs in the process of adding $k_{\text{base 2}}$ and $n_{\text{base 2}}$. For an example, take $k = 3$ and $n = 4$. In the binary addition of k and n there is no carry (see table below on the left). Thus, $\binom{3+4}{3}$ is not divisible by 2. In fact, $\binom{7}{3} = 35$. Now change n to 5 (see table on the right). Here a carry occurs in the base 2 addition at all three rightmost digits. Thus, according to Kummer, $\binom{3+5}{3}$ is divisible by 2. In fact, $\binom{8}{3} = 56$.

	decimal	binary
k	3	11
n	4	+100
$k+n$	7	111

	decimal	binary
k	3	11
n	5	+101
$k+n$	8	1000

1. Using the same method determine whether the following combinations are divisible by 2. Verify your results by computing the numerical values of these combinations.

   a. $\binom{9}{6}$ b. $\binom{23}{16}$ c. $\binom{21}{10}$

2. Using Kummer's criterion determine whether the following combinations are divisible by 3. Verify your results by checking the numerical values of these combinations. Hint: Convert the numbers k and n in base 3, then add them and find out whether a carry occurs.

   a. $\binom{9}{6}$ b. $\binom{23}{16}$ c. $\binom{21}{10}$

7.6B

2 divides if a "carry" occurs in the process of adding $k_{\text{base 2}}$ and $n_{\text{base 2}}$.

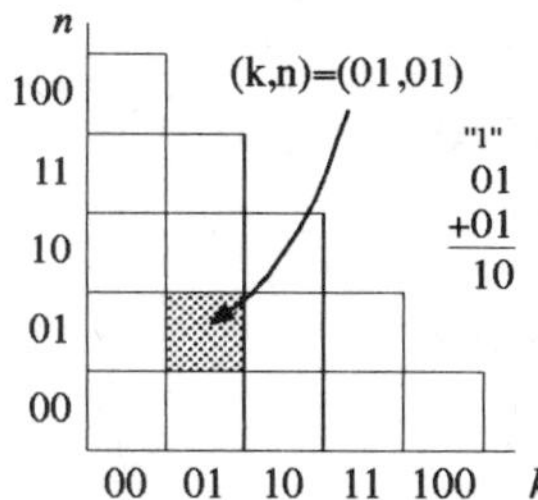

3 divides if a "carry" occurs in the process of adding $k_{\text{base 3}}$ and $n_{\text{base 3}}$.

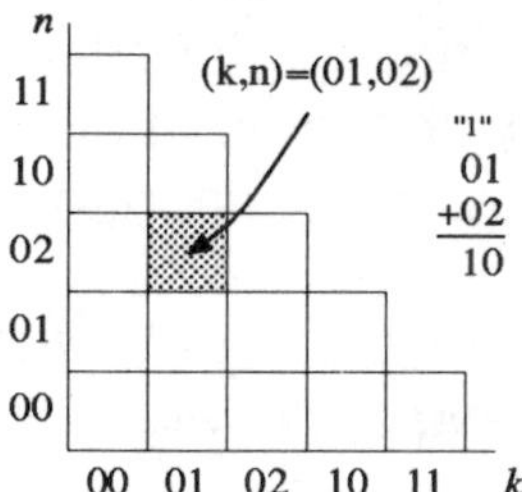

3. Find another cell in the above grid that contains a combination divisible by 2.
4. Find another cell in the above grid that contains a combination divisible by 3.
5. The array of the combinations from Pascal's triangle has been copied below. Shade each cell that is not divisible by two.

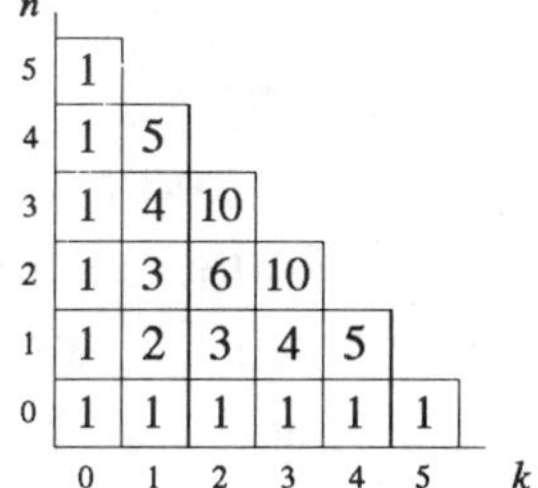

6. Shade each cell in the empty grid below for which addition of the binary coordinates k and n does not involve a "carry" during the addition.

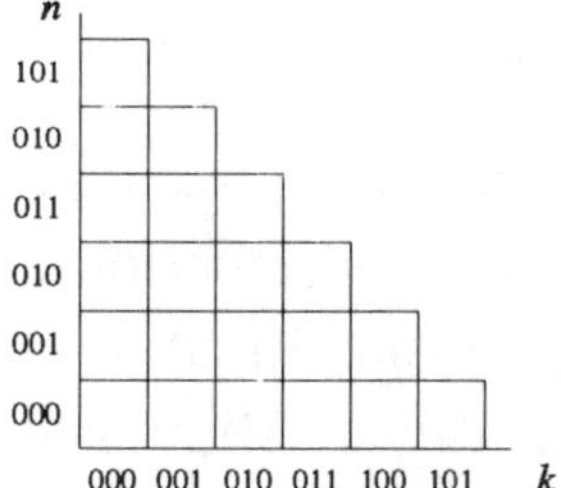

Each of the coloring patterns initiated above could be continued through many additional rows. However, without the aid of Kummer's divisibility criterion for determining whether or not a cell should be colored, successive rows of Pascal's Triangle must be computed. While each value could be computed by adding the two appropriate values in the preceding row or by evaluating combinations, either approach becomes increasingly cumbersome. Kummer's Criterion offers a simple addition test for any cell based upon that cell's address.

Now we can answer the question raised at the beginning of this activity. Let us look at the entry in row n and column k of Pascal's triangle. It is the combination $\binom{k+n}{k}$, and it is shaded if this number is divisible by 2. By Kummer's criterion this means that there occurs at least one carry in the binary addition of k and n. The first such carry occurs *when corresponding binary digits of k and n are both equal to 1.* But this is precisely the address-derived characterization of the shaded cells in the Sierpinski triangle! So Kummer's observation really explains the connection between the two methods to generate the Sierpinski pattern.

7.7 MAPPINGS

7.7A

An affine transformation maps one figure to another in a one-to-one fashion. An affine transformation is called a similarity when the image figure is a proportionally reduced or expanded copy of the original. In such a case the ratio r of expansion or reduction is the same for all mapped dimensions.

1. For each of the given original figures, draw the image figure using the given ratio r. Label the dimensions of the image figure as determined by the ratio r.

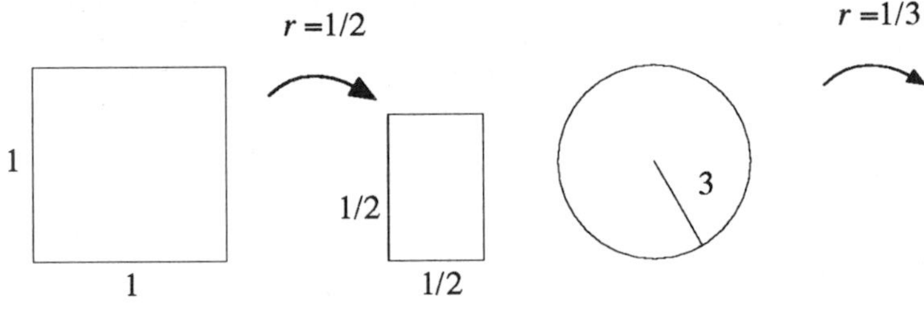

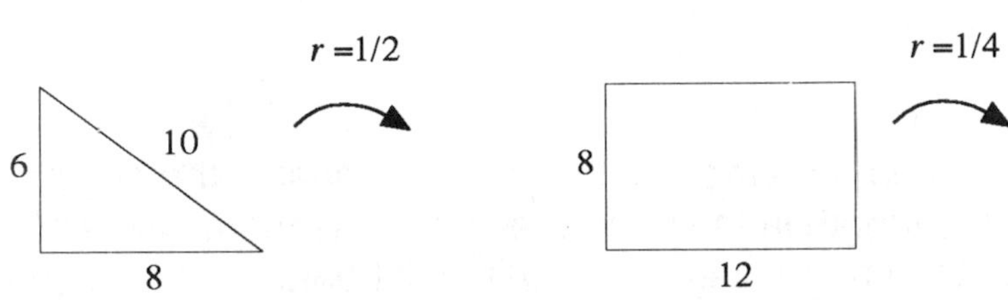

2. Map the initial figure using a similarity transformation with ratio r into a region having the same dimensions as the original figure a sufficient number of times so as to just cover the region without overlapping images.

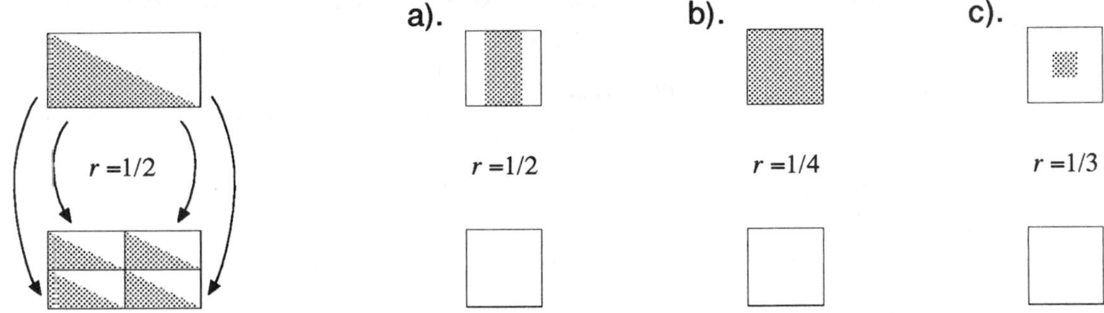

An affine transformation is a similarity when the ratio r of expansion or reduction is the same for all mapped dimensions. Additionally a rotation, translation, or reflection may be applied. A simple class of affine mappings different from similarities is given by transformations that multiply length by a constant ratio r in one direction and by a different constant s in another direction.

7.7B

3. For each of the given original figures, draw the image figure using the given ratios r for horizontal lengths and s for vertical distances. Label the dimensions of the image figure as determined by the ratio r.

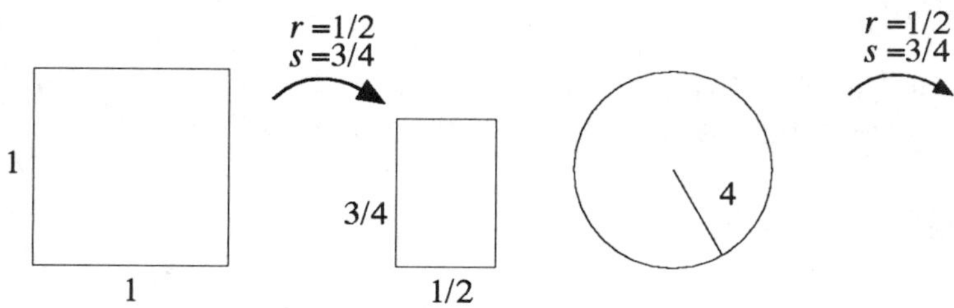

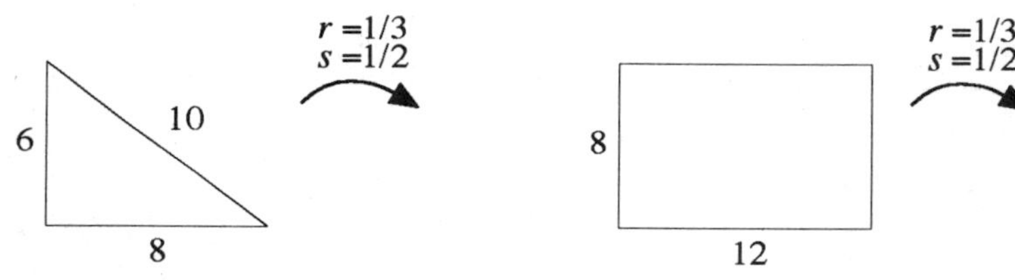

4. In the top left example above, compute the length of the diagonal in the original square as well as the diagonal length in the image rectangle. Is the multiplying ratio from the original diagonal to the image diagonal equal to the value of ratio r or ratio s? If not, what is the ratio?

5. Map the initial figure using an affine transformation with horizontal ratio r and vertical ratio s into a region having the same dimensions as the original figure a sufficient number of times so as to just cover the region without overlapping the images.

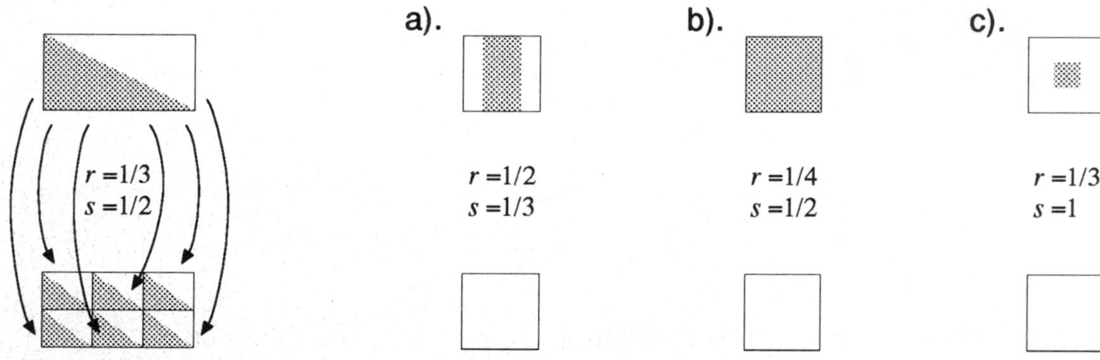

In Acitivity 7.1 Templates A and B each represented a similarity transformation from Illustrations A and B with $r = 1/2$. Likewise, the mappings discussed in Activities 7.2 and 7.3 were similarity transformations. In the next Activity 7.8, the mappings are affine mappings in which the horizontal and vertical ratios differ.

7.8 COPYING MACHINES GONE WILD

7.8A

A mapping used in the process of transferring a small copy of an original figure in a TV studio to a location k on a receiving TV screen was denoted in prior activities by w_k. Until now copies were reproduced without distortion. Now suppose that the transmission process systematically elongates some of the images as shown in the diagram to the right.

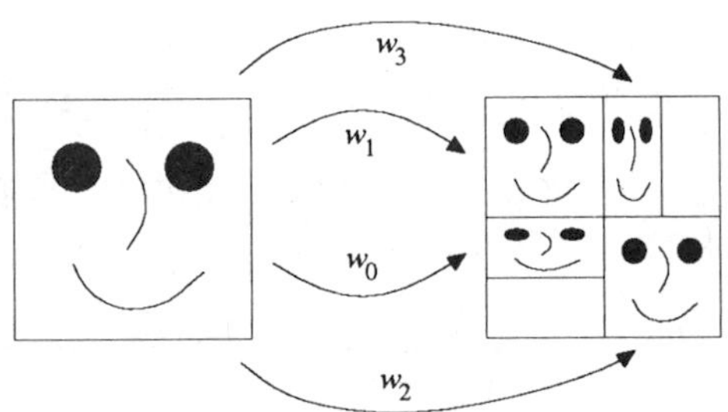

1. Use a duplicating machine to make a copy of page 7.8E. Cut the copy along the vertical dotted line in order to separate the right hand section entitled Transformed Copies using w_k's.
2. Cut out the four STAGE 0 COPIES from the strip of Transformed Copies. Paste each of the pieces corresponding to the mappings w_0, w_1, w_2, and w_3, into sections 0, 1, 2, and 3, respectively, of the IMAGE 1 grid on page 7.8E. The result should look like the right hand image shown in the picture above.
3. Cut out the four STAGE 1 COPIES from the strip of Transformed Copies. Paste each of the pieces corresponding to the mappings w_0, w_1, w_2, and w_3, into sections 0, 1, 2, and 3, respectively, of the IMAGE 2 grid on page 7.8E.
4. Repeat the cut and paste process practiced in problems 2 and 3 above using the four STAGE 2 COPIES from the strip of Transformed Copies and the IMAGE 3 grid on page 7.8E.

Instead of transmitting a "Smiley Face", suppose that a television studio uses the w_0, w_1, w_2, and w_3 mappings to transmit copies of the X shown on the STAGE 0 screen below.

5. Draw the result that would appear on the STAGE 1 screen. Then draw the STAGE 2 screen that would result if the television studio subsequently transmitted copies of the STAGE 1 result.

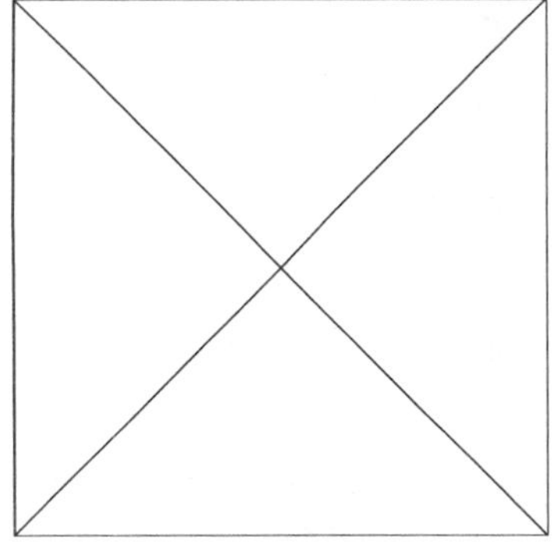

Stage 0

Stage 1

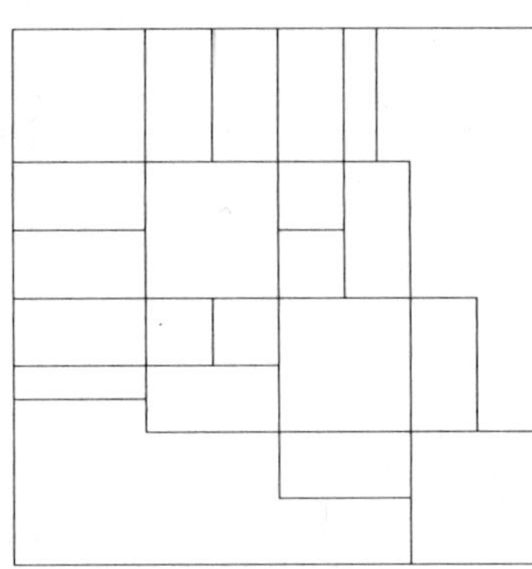

Stage 2

7.8B

6. If the transmission process were repeated many times, would the process lead to a different final image for the "Smiley Face" as compared to the final image generated by the X? Explain your answer.

Now suppose that a transmitting TV studio uses the Chaos Game discussed in the prior activities to generate a sequence of dots on a TV screen.

i. Arbitrarily select a point anywhere on the screen.

ii. Randomly select one of the four mappings w_0, w_1, w_2, or w_3.

iii. Draw a new dot on the screen in the small region identified by the mapping chosen in step (ii) above. Place the new dot in the small region so that its position is proportionately the same as the location of the prior dot on the entire screen.

iv. Using the last dot placed repeat steps (ii) and (iii).

The illustration below relates to the random sequence of mappings $\cdots w_0 w_2 w_3$ (IMPLEMENTED RIGHT TO LEFT).

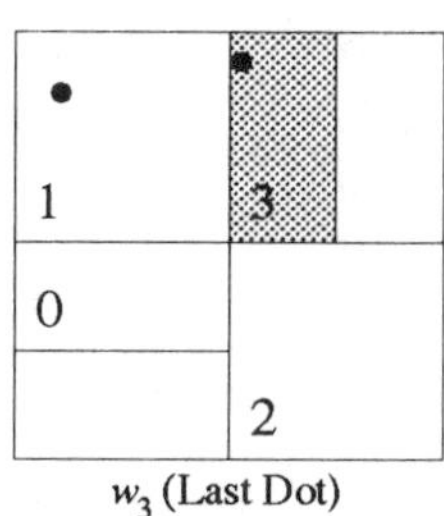

w_3 (Last Dot)

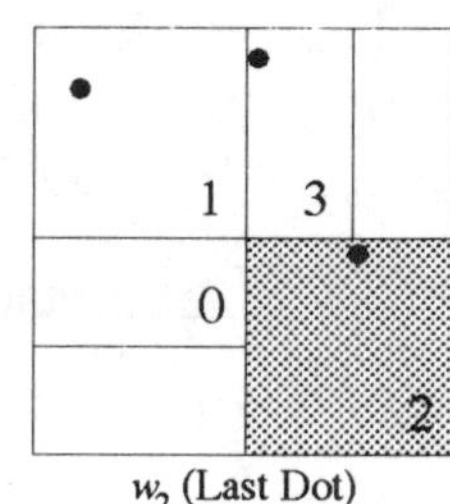

w_2 (Last Dot)

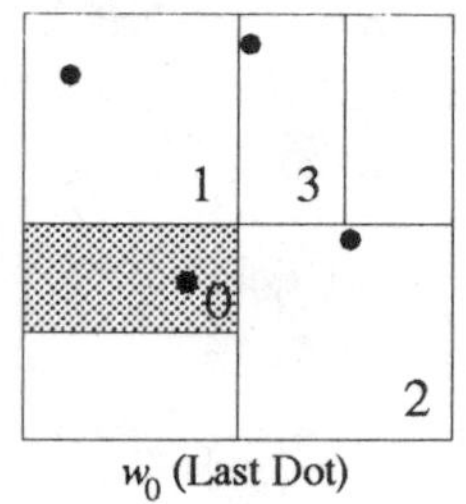

w_0 (Last Dot)

7. Apply steps (i)–(iv) of the chaos game beginning with the dot shown on the TV screen below.

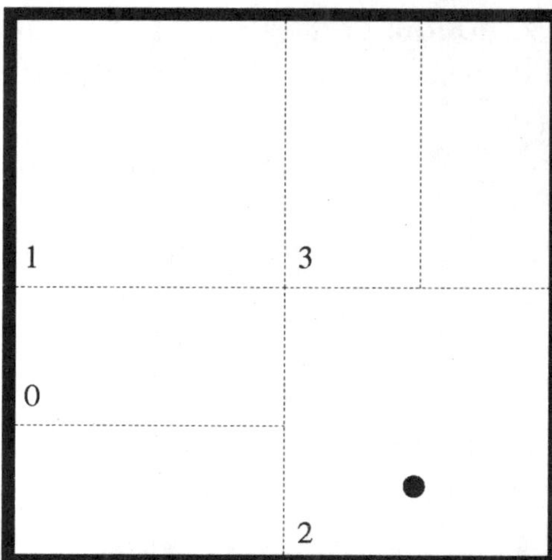

7.8C

The following calculator programs can be used to continue the iteration process initiated in problem 5 above. Key the appropriate program into your graphing calculator.

CASIO

```
Range 0,1.33,.1,0,1,.1
0 → N
.5 → P
.5 → Q
Lbl 1
Plot P,Q
N+1 → N
N>2000 => Goto 6
Ran# → X
X<.3333 => Goto 2
X<.5 => Goto 3
X<.8333 => Goto 4
Lbl 2
.5P → P
.5Q+.5 → Q
Goto 1
Lbl 3
.5P → P
.25Q+.25 → Q
Goto 1
Lbl 4
.5P+.5 → P
.5Q → Q
Goto 1
Lbl 5
.25P+.5 → P
.5Q+.5 → Q
Goto 1
Lbl 6
```

TEXAS INSTRUMENTS

```
:ClrDraw
:0 → Xmin
:1.33 → Xmax
:0 → Ymin
:1 → Ymax
:0 → N
:.5 → P
:.5 → Q
:Lbl 1
:Pt-On(P,Q)
:N+1 → N
:If N>2000
:End
:rand → X
:If X<.3333
:Goto 2
:If X<.5
:Goto 3
:If X<.8333
:Goto 4
:Goto 5
:Lbl 2
:.5P → P
:.5Q+.5 → Q
:Goto 1
:Lbl 3
:.5P → P
:.25Q+.25 → Q
:Goto 1
:Lbl 4
:.5P+.5 → P
:.5Q → Q
:Goto 1
:Lbl 5
:.25P+.5 → P
:.5Q+.5 → Q
:Goto 1
:
```

7.8D

The sequence of four images below displays the result of continuing the chaos game for a few thousands of points.

8. Discuss the sense in which the whole image can be partitioned into parts in such a way that each of these parts is a replica of the whole image. Show how the parts can themselves be partitioned into sections that possess this same self-similarity property.

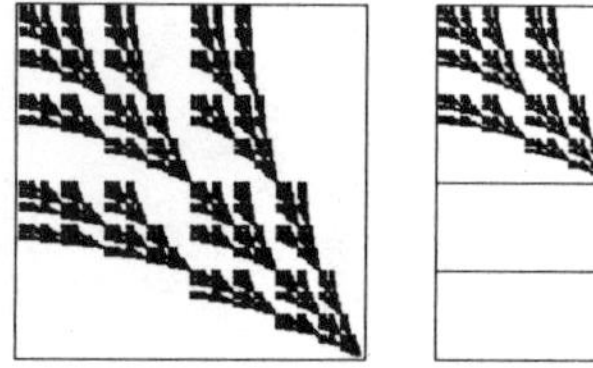
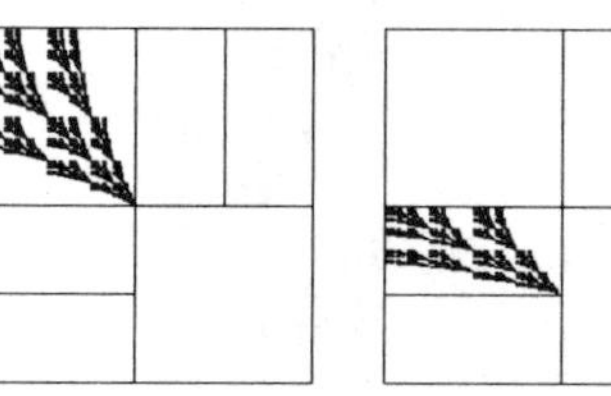
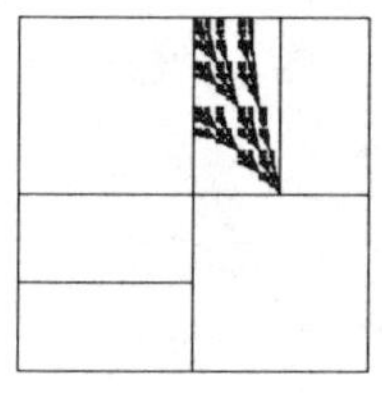
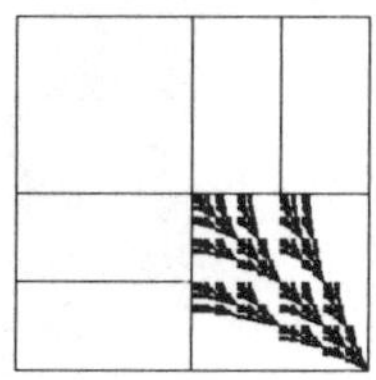

9. How does knowing the nature of the mappings w_0, w_1, w_2, and w_3 give a kind of geometric genetic code for the resulting image?

7.8E

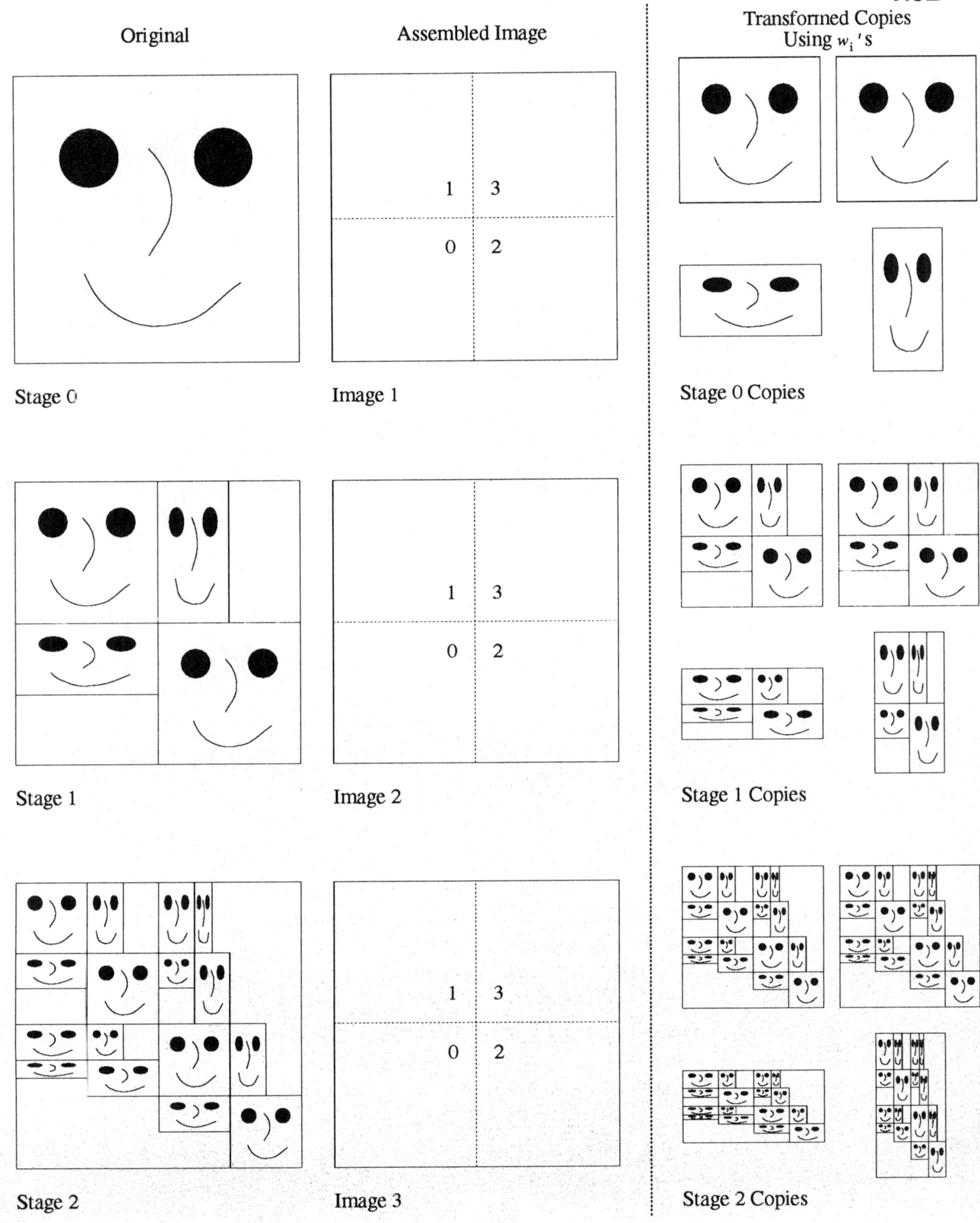
Original
Assembled Image
Transformed Copies Using w_i's
1
3
0
2
Stage 0
Image 1
Stage 0 Copies
Stage 1
Image 2
Stage 1 Copies
Stage 2
Image 3
Stage 2 Copies

7.9 THE SIERPINSKI TRIANGLE, SIERPINSKI CURVE, AND SELF-SIMILARITY 7.9A

In prior Activities we have used the notation w_k to denote mappings of some figure back onto a subregion of itself. Imagine repeatedly applying three mappings w_1, w_2, and w_3 to a triangle as shown to the right. The sequence of images that appears approaches the Sierpinski Triangle as the final limit figure.

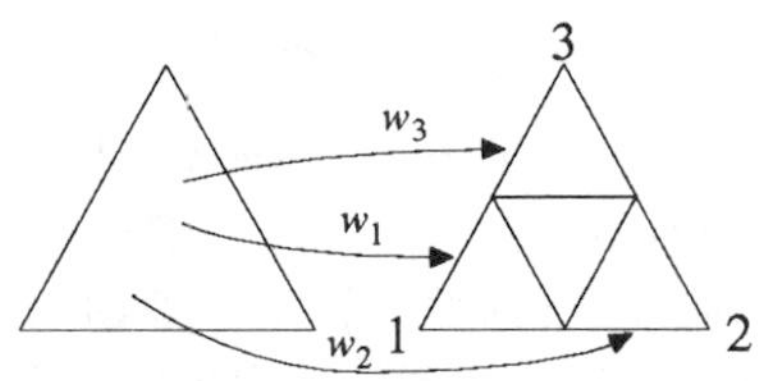

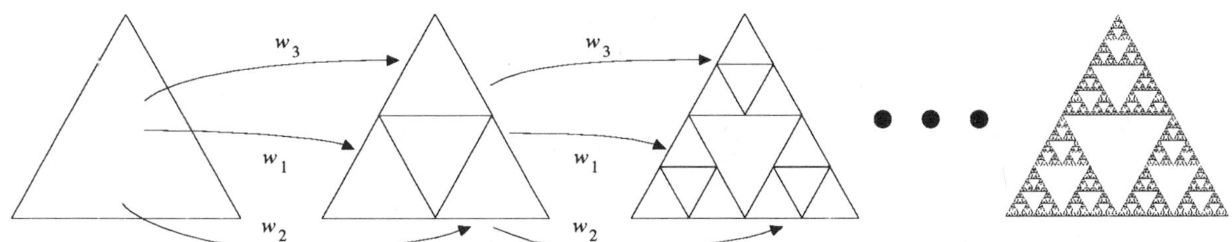

1. Cut out the three small copies of the completed Sierpinski Triangle from down below. Each copy is a one-half size replica of the Sierpinski Triangle shown below. Place one copy into each of the three numbered sections on the following SIERPINSKI TEMPLATE.

2. Into how many distinct new orientations can a small copy be rotated so as to exactly fit within a given region on the template?

Sierpinski Triangle

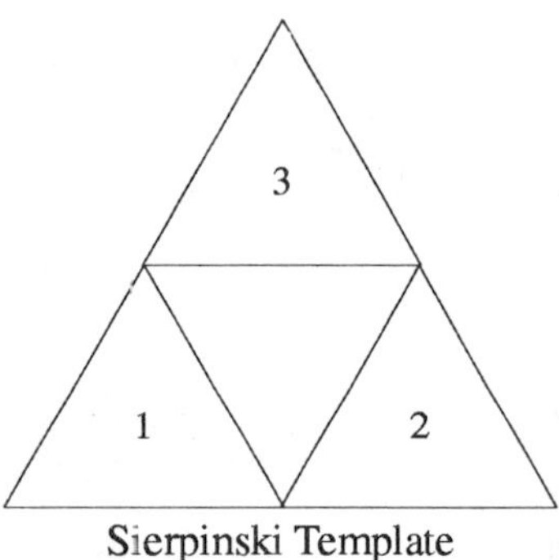

Sierpinski Template

3. In view of your answer to question 2, how many distinct rotations (including rotations of 0 degrees) exist that map a small copy of the Sierpinski Triangle into region 1?

4. If a small copy of the Sierpinski Triangle can be pasted into any one of three distinct rotations within a given region (no flips allowed), explain why the Sierpinski Triangle can be assembled in 27 distinct ways by means of mappings.

Fractal objects typically possess a property referred to as self-similarity. When an object is self-similar, it can be partitioned into parts that together comprise the whole object in such a way that each of these parts is a replica of the whole object.

Figures for question 1

7.9B

5. State the self-similarity property in terms of mappings.

The illustration below suggests the Sierpinski Triangle still results even when the mappings w_1, w_2, and w_3 rotate their small copies into any one of the three permissible orientations. In this example, w_3 performs a 0 degree counterclockwise rotation while w_1 and w_2 both perform 120 degree counterclockwise rotations.

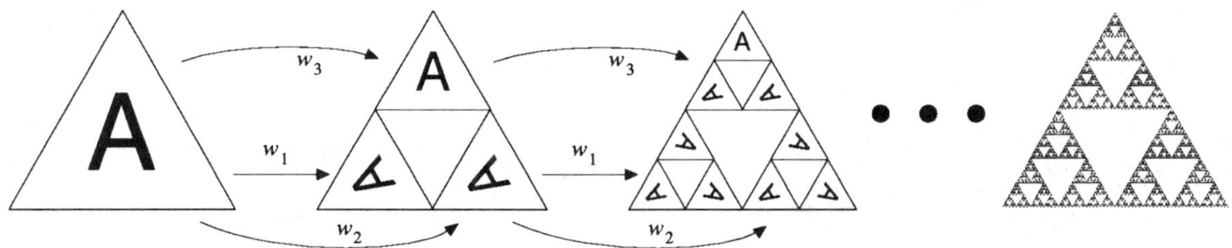

6. Suppose that the mappings w_1, w_2, and w_3 include counterclockwise rotations through 240, 120, and 0 degrees, respectively. Complete the Stage 1 and Stage 2 images shown below if the original Stage 0 triangle contains the letter "J".

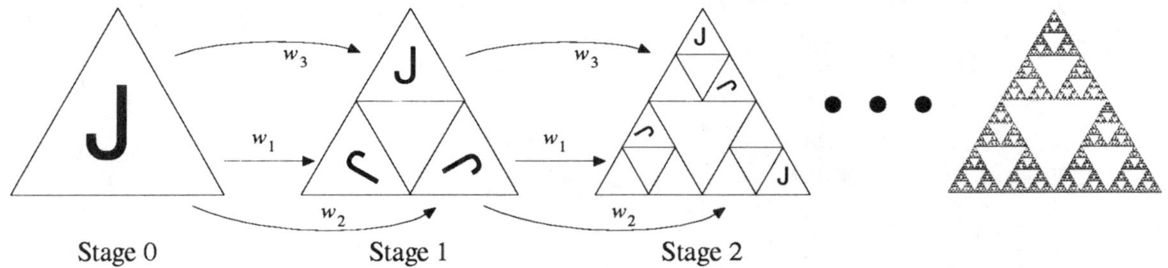

7. Explain why adding designs (such as the letter A or J) to the original figure has no bearing upon the eventual outcome.

8. Again suppose that the mappings w_1, w_2, and w_3 include counterclockwise rotations through 240, 120, and 0 degrees, respectively. Finish tracing the Stage 3 and Stage 4 Sierpinski Curves on the triangular lattices provided by the successive stages of the Sierpinski Triangle construction.

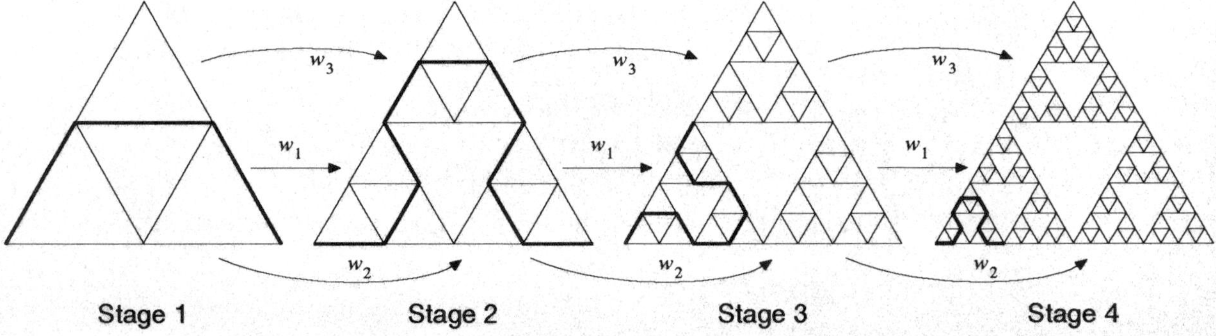

9. What is the final limit figure for the Sierpinski Curve?

10. Explain why the Sierpinski Curve is self-similar.

Unit 8
IFS and Geometric Genetic Codes

KEY OBJECTIVES, NOTIONS, and CONNECTIONS

Activities contained in this unit discuss the relations between geometric fractal shapes and their counterparts, their mathematical descriptions. These formal definitions of the fractal shapes make use of important basic mathematical concepts: linear matrix algebra, corresponding transformation equations, and the group of square isometries. These concepts are explored using a small family of fractals derived from the Sierpinski triangle. The characteristic features of such a fractal can be distinguished by a short formal description, the geometric genetic code. This unit highlights the connections between geometry and algebra showing how each can help to better understand the other.

Connections to the Curriculum

The material covered in these strategic activities integrally relate to content already included in contemporary mathematics programs. Consequently, these activities can be invested into the curriculum as stand alone units, or they may be periodically integrated into existing content sequences as applications or enrichment activities.

PRIMARY CONNECTIONS:

Mappings	Function Composition
Geometric Transformations	Visualization
Convergence	Linear Transformations
Geometric Patterns	Iteration
Symmetry	Similarity

SECONDARY CONNECTIONS:

Counting Techniques	Matrices
Coordinate Geometry	Groups
Graphing Calculator	

Underlying Notions

Abelian Group

When a set S of elements combine under some operation $\circ$ in such a way that

1. operation $\circ$ is closed, associative, and commutative;
2. S contains an identity element for the operation $\circ$;
3. for each element in S there exists a unique inverse element in S;

then the elements of S form an abelian group under operation $\circ$.

Geometric Genetic Code The underlying self-similarity structure of a fractal image implies that a set of transformations exist that map reduced copies of the fractal to its parts. Since these transformations both describe the self- similarity structure as well as the process that generates the image, the mappings together provide a kind of geometric genetic code for the figure.

Transformation When points in one geometric figure are mapped to the points of another figure, the mapping is referred to as a transformation.

Iterated Function System In an iterated function system more than one transformation repeatedly acts upon points or figures to produce a sequence of image points or image figures.

MATHEMATICAL BACKGROUND

The Bigger Picture

Benoit Mandelbrot, sometimes called "The Father of Fractals," observes that many things in nature are fractal especially in regards to possessing the fundamental feature of self-similarity. This underlying self-similarity structure present in so many natural phenomena or artifacts provides a key to simulating the object or natural process in a computer. The presence of this property implies that the aritifact can be partitioned into parts in such a way that each part is an image of the whole object under a unique mapping. Taken together, these mappings that impress copies of the whole onto its component parts exist as a kind of genetic code of the self-similarity structure imbedded within the whole structure. Moreover, the ability to unlock the genetic codes of distinct artifacts permits a comparison of one to another by reference to relationships observable between the respective genetic codes.

Collectively, the set of mappings that impress copies of the whole figure onto its component parts form an Iterated Function System. A construction process that utilizes such an Iterated Function System typically involves two elements:

(i) a kind of template that supplies a geometric blueprint for the self-similarity structure, and

(ii) a set of transformations from which some are selected as the specific mappings that copy reduced replicas of the whole image into the various sections of the blueprint/template.

In this Unit we will use a square divided into four quadrants as our template, and three of the four quadrants will provide a blueprint for the self-similarity structure. We will select mappings from a set known as the "dihedral group of the square," a set of mappings derived from the symmetries of the square as described below.

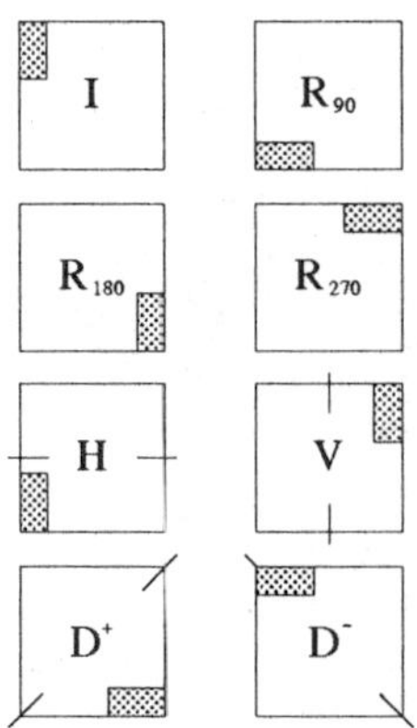

Four lines divide a square into symmetric halves. By regarding any one of these lines as an axis, flipping the square about such an axis causes the points on one side of the square to be reflected onto the points of the other half. This kind of reflection of the square about an axis of symmetry represents a mapping of the square back onto itself. Since the square possesses four lines of symmetry, as shown in the illustration to the left, four mappings H, V, D^+, and D^- exist that reflect the square back onto itself. In addition to these four reflections, we can rotate the square back onto itself by turning it counterclockwise through multiples of 90 degrees about its center point. Three rotations through 90, 180, and 270 degrees effectively map the square back onto itself by this turning point symmetry. Finally, a unique identity transformation I exists that neither rotates nor flips the square but which nonetheless maps the points of the square identically back onto themselves. Accordingly, the eight rigid motions of the square include four reflections, three rotations, and one identity transformation.

Additional Readings

Chapter 5 of *ChaosandFractals*, H.-O. Peitgen, H. Jürgens, D. Saupe, Springer-Verlag, New York, 1992.

USING THE ACTIVITY SHEETS

8.1 Symmetries

Specific Directions. When points in one geometric figure are mapped to the points of another figure, the mapping is referred to as a transformation. Imagine the possibility that the points of the original figure are mapped to points contained within the same figure. If each point is mapped to itself then the transformation is commonly called an identity mapping and it is often denoted by I. Sometimes a figure can be mapped onto itself without distortion by a transformation other than the identity transformation. In such a case the mapping is a rigid motion which either rotates or reflects the figure back onto itself. In the case of a square, these motions include three counterclockwise rotations through 90, 180, or 270 degrees and four reflections through lines which bisect the square into two equal and symmetric halves. The objective of this activity is to practice identifying the rotational and line symmetries present in a variety of figures.

Implicit Discoveries. The key to determining the set of transformations that map a figure rigidly back onto itself is found in the set of rotational and line symmetries of the figure.

8.2 Compositions

Specific Directions. In this activity we consider the set of rigid motions that map the points of a square back onto the points of the same square without distortion. When two of these transformations are applied to the square in succession, the combined result of the two mappings is called a composition. Since only eight motions exist which map the square back onto itself in the prescribed fashion, the result of any composition of two or more motions must be equivalent to one of these eight possible single motions. For example, a 270 degree rotation of the square generates in a single motion the same result that could be produced by the composition of a 90 degree rotation and a 180 degree rotation.

The ***symmetry computer*** introduced in this activity is a simple but effective hands-on aid for computing arbitrary compositions of the eight rotations and flips of the square. To begin make a photocopy of the manipulative at the bottom of Activity 8.2C and cut out the left part of the figure, a square with the top left corner removed. Placed on top of the other figure the cutout reveals the starting position of the symmetry computer. This is the I (identity) transformation, which we can interpret as the initial orientation of a square before any operations are carried out. When considering a sequence of rigid motions of the square, simply apply them to the lose piece of the computer, and place it back onto the template. The cut out corner will show the result of the operation. For example, after a rotation of 90 degrees (rotate always counterclockwise) the symbol R_{90} is visible. Apply a second motion, for example, the reflection at the diagonal, D^+. The result will be H, i.e., the composition in total equals the horizontal reflection H.

Implicit Discoveries. The table of all possible pairwise compositions of the eight motions of the square back onto itself forms a transformation group.

Extensions. Identify the six symmetries of an equilateral triangle. Then construct and discuss the table for the 36 possible compositions of two such symmetry transformations.

8.3 Algebraic Mappings

Specific Directions. From a geometrical point of view, a mapping assigns to each point (x, y) in a plane some unique image point (x', y'). When a systematic rule is used to effectively implement such a mapping, the specific rule can usually be represented algebraically as a system of two equations. In these Unit 8 activities, the mappings are one-to-one, linear transformations. The equations of such transformations as these are themselves linear equations of the form

$$x' = ax + by,$$
$$y' = cx + dy.$$

Once the coefficients a, b, c, and d of these equations are known, the associated transformation is also completely determined. Consequently, matrices provide an economical way of representing linear transformations since we can conveniently collect the coefficients into a coefficient matrix as shown in the diagrams below.

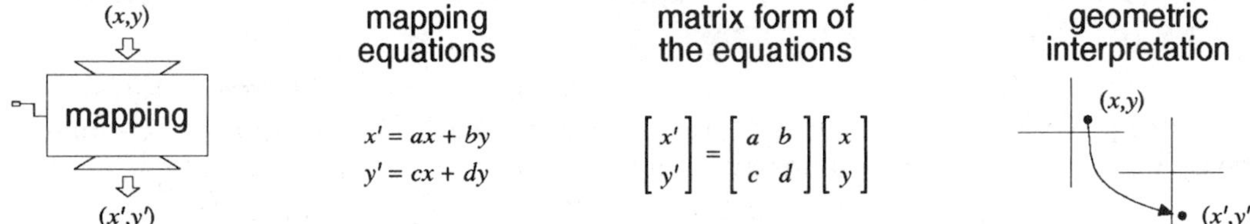

Implicit Discoveries. For each of the eight motions of the square back onto itself, a unique algebraic system of two equations exists that can be used to compute the images of points forming the square.

Extensions. Explore the algebraic systems of two equations that exist for each of the six symmetries of an equilateral triangle.

8.4 Fractal Images

Specific Directions. Fractals objects exhibit an underlying and fundamental self-similarity structure. Objects with this property can be partitioned into parts that together form the whole object, but in such a way that each of these subsection is a replica of the whole. Since these strategic parts are smaller than the whole, each one must be a reduced but possibly reoriented copy when compared to the orientation of the whole object itself. Consequently, identifying the subsections that give form to the self-similarity structure can be complicated by the various reorientations of the subsections. In this activity we restrict ourselves to the eight possible reorientations represented by the rotations and reflections of a square as practiced in Activities 8.1, 8.2, and 8.3.

Implicit Discoveries. Rotations and reflections are symmetries but they are what produce the self-similarity in these fractal structures. It follows that self-similarity can be viewed as a special kind of symmetry.

Extensions. Explore the effect of performing the six symmetries of an equilateral triangle on the reduced copies when generating the Sierpinski triangle in an equilateral triangle template.

8.5 Building Fractals

Specific Directions. Instead of identifying the fundamental parts that provide the inherent basis for the self-similarity structure of an already existing fractal object as practiced in Activity 8.4, the objective in this activity is to prescribe the relative positions between the fundamental parts and to determine the orientation of these parts in the whole so as to build our own "designer" fractals.

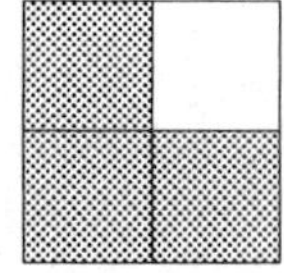
Template

Begin by dividing a square into four quadrants each of which represents a possible location for a reduced copy of the eventual entire figure. This square will serve as a template for the following construction process. Assume that the underlying self-similarity structure of the eventual image will issue from three basic parts. Select the top left, lower left, and lower right quadrants of the square for these three parts. Choose an orientation for each part from among the eight possible transformations represented by the rotations and reflections of a square practiced in Activities 8.1, 8.2, and 8.3. Now, starting with an original square containing some arbitrary figure, make three reduced replicas and in turn place each within one of the three selected quadrants of the template. Be sure to orient each according to the specific transformation that you have assigned to the three selected quadrants. The resulting image thus formed on the template provides the new figure for the reduction, replication, and rebuilding process. The limit figure that appears from continually repeating this process is the fractal encoded by the template and the three selected transformations.

Activity 8.5D is designed to show the developing fractal image for the code (I, R_{180}, R_{90}) as it moves from stage to stage. The sheet contains 30 squares showing stage-4 images. Apply the code to combine 3 of these stage-4 images to construct a stage-5 structure. Apply the code again as you combine 3 of the stage-5 images to construct a stage-6 structure. Combine 3 of the stage-6 images to construct a stage-7 structure. Watch how the detail grows.

Since no scaling occurs from stage to stage, one must assume larger and larger initial squares as the building process progresses. Consider each stage-4 square as a 1×1 unit square. The following table shows the corresponding sizes for successive stages.

Stage	4	5	6	7
Initial Square	1×1	2×2	4×4	8×8

Activities 8.5E and F repeat the process for the fractals with codes of (R_{90}, I, R_{270}) and (R_{180}, R_{270}, I). However, here a final image is given as each initial square. As one builds a larger and larger structure, the built in self-similarity at different scales becomes more and more visible.

Other repeated images can be used to build additional fractal structures by this method. There are 64 with codes that make use of the symmetries of $I, R_{90}, R_{180}, R_{270}$. Reflections should be avoided since both sides of the building squares would then have to be used.

Implicit Discoveries. Any of the 64 rotation combinations possible from $I, R_{90}, R_{180}, R_{270}$ can be used to construct successive stages of a developing fractal as was done in questions 13–16 with (I, R_{180}, R_{90}).

Extensions. Another approach similar to that done in questions 13–16 would be to start with repeated images of the limit figure itself, building a larger and larger image to show better the nature of the complexity of these fractal structures. In doing this pasting up activity, be sure that it is understood that the fractal itself is not growing. Rather, the initial stage is assumed to be larger and larger as more parts are assembled.

8.6 Summarizing the Process

Specific Directions. Activity 8.4 introduced the geometric genetic code and Activity 8.5 showed how this code, along with the appropriate template, can be used to build the fractal. The geometric genetic code alone reveals the transformations of the copies once they are properly reduced and located in the square grid. This activity looks at a table format that lays out more fully the underlying mechnism of the complete iterated function system operating within the square. It addresses all three aspects of the process, the reducing, the replication and the building. The key elements in this encoding include identifying the base, numbering of the copies, the transformation imposed on them, telling how each copy is scaled, and where the transformed and reduced copies are finally placed. These entries in the table below establish the characteristics found in the Sierpinski trangle, as generated within a square template. All examples in this activity follow this format.

Base	Copy	Orient-ation	Scale	Offset	
				Hori	Vert
	1	I	1/2	-1/2	1/2
2	2	I	1/2	-1/2	-1/2
	3	I	1/2	1/2	-1/2

The base of 2 means the sides of the square are divided into two equal parts. The orientation identifies the symmetry of the square that is applied to that copy. The scale of 1/2 reduces each copy to the size of a quadrant. The horizontal and vertical offsets identify the symmetry of the square that is applied to that copy.

To help understand the role of the offset entries, think first of the reduction as a dilation towards the center. The offsets move the square from that position.

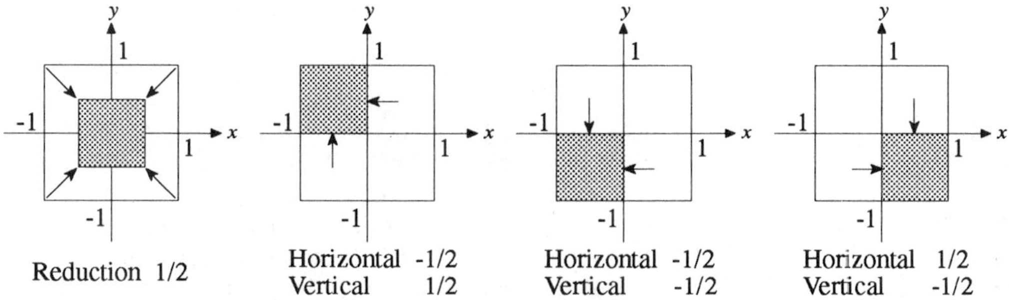

Matrices can be used to analytically summarize the combined effects of re-orienting, scaling, and translating. The equation

$$\begin{pmatrix} x' \\ y' \end{pmatrix} = s \cdot \mathbf{M} \begin{pmatrix} x \\ y \end{pmatrix} + \begin{pmatrix} h \\ v \end{pmatrix}$$

identifies an image $\begin{pmatrix} x' \\ y' \end{pmatrix}$ for each original $\begin{pmatrix} x \\ y \end{pmatrix}$. The particular matrix **M** corresponds to a special symmetry of the square, s is the scaling factor, and h, v determine the horizontal and vertical offsets, respectively.

Implicit Discoveries. When using a 2×2 grid, since all reductions are 1/2 size and $N = 3$ copies are always placed in three of the four quadrants of the square, all structures so generated are closely related to the Sierpinski triangle. Each has a fractal dimension of about 1.58.

$$D = \log N / \log(1/s) = \log 3 / \log 2 \approx 1.58$$

Extensions. Explore fractals generated in a base 3 grid, where the reduction scale is 1/3 and offsets are $-2/3, 0, 2/3$ on the initial square used throughout this activity. One special example is the Sierpinski carpet, where 8 reduced copies are placed around the outside positions of the 3×3 grid. Encode the process in an IFS table and show that the fractal dimension is about 2.79.

8.7 Chaos Game Variations

Specific Directions. Activity 8.5 introduces an algorithm for generating a fractal image that starts from a square containing an arbitrary figure. After repeatedly reducing, replicating, and rebuilding the figure at each stage of the construction process, a fractal image emerges as the limiting image of the process.

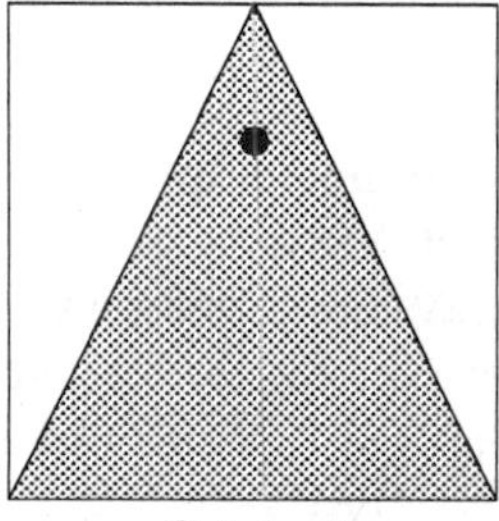
Original

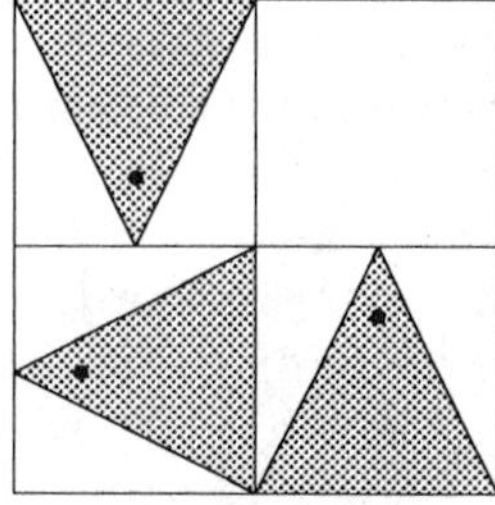
Image Sheet 1

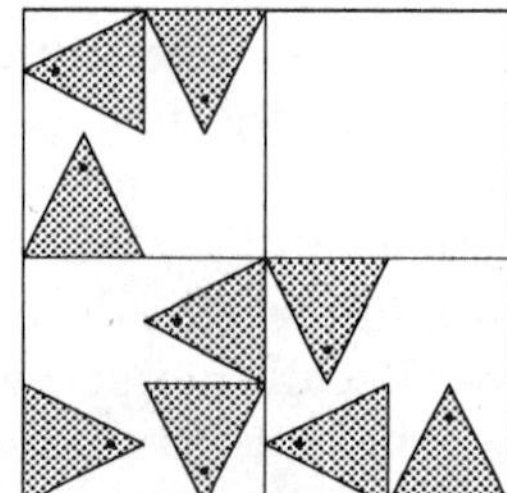
Image Sheet 2

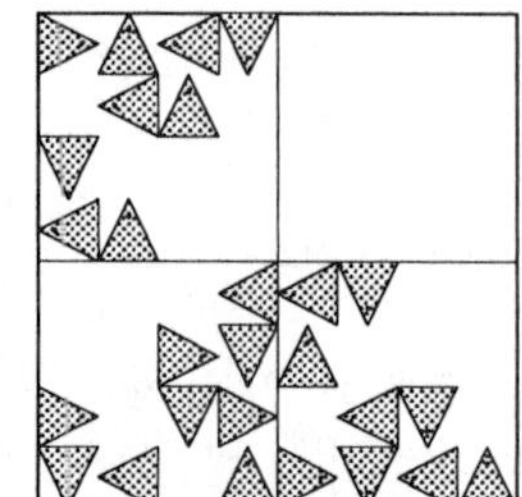
Image Sheet 3

The initial figure is rather arbitrary since, in repeating the reduction process, even a large triangle shrinks to the size of a small dot. This fact suggests that a chaos game could be employed to generate the same image as that produced by the reduce, replicate, and rebuild algorithm in Activity 8.5.

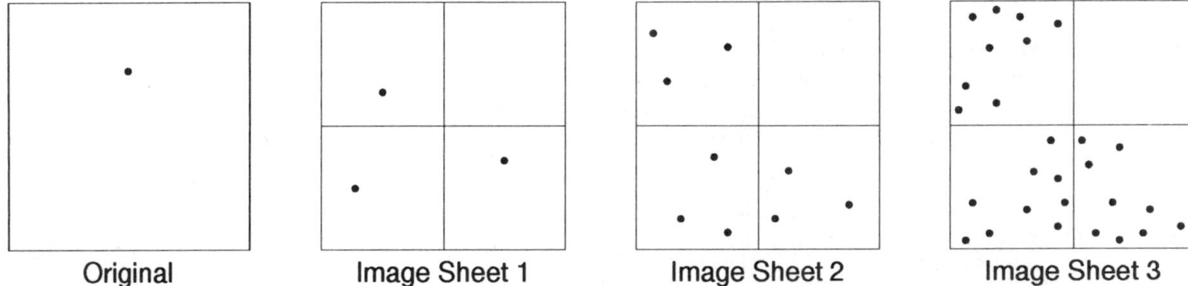

As in Activity 8.5, begin by dividing a square into four quadrants each of which represents a possible location for a reduced copy of the eventual entire figure. This square will serve as a template for the following construction process. Assume that the underlying self- similarity structure of the eventual image will issue from three basic parts. Select the top left, lower left, and lower right quadrants of the square for these three parts. Choose an orientation for each part from among the eight possible transformations represented by the rotations and reflections of a square practiced in Activities 8.1, 8.2, and 8.3. Now, starting with an original square containing an arbitrary interior point, randomly select one of the three quadrants on the template. Map the point into the same relative position within the selected quadrant as that occupied by the original point within the large square. Be sure to orient the new point within the quadrant according to the specific transformation that you have assigned to that quadrant. The new point is now not only within a selected quadrant, but it also lies within the boundaries of the large square. Continue to randomly select one of the three quadrants, and repeatedly copy the location of the last mapped point from within the large square into the newly selected quadrant. Always be sure to orient the new point within the quadrant according to the specific transformation associated with that subregion. The limit figure that appears from continually repeating this process is the fractal encoded by the template and the three selected transformations. Specifically, the resulting image depends completely upon the configuration of the template and the transformations selected for each subregion.

Implicit Discoveries. Once the template is known, the transformations supply a kind of geometric genetic code for the resulting self-similarity structure that appears, and indeed, for the very construction process itself.

8.8 Families of Fractals

Specific Directions. Activity 8.1 investigated the eight symmetries of a square based upon 1) rotations about the center point and, 2) lines of reflection that divide the square into two symmetric halves. In subsequent activities these eight symmetries were treated as geometric transformations and each of the eight possibilities was formulated algebraically in terms of a set of linear equations. Activities 8.4 to 8.7 demonstrated how these mappings could be used to create fractal images.

Some of these fractal images exhibit line symmetries themselves. For these fractals, more than one combination of three mappings, each selected from among the eight possiblilities, all produce the same image. Consequently, the presence of a line symmetry in the fractal image implies that the geometric genetic code is not unique. When a line symmetry exists in a fractal image, the self-similarity structure of the fractal provides the key for determining the distinct geometric genetic codes that describe the figure. Recall that self-similarity implies that the figure can be decomposed into distinct parts each of which is a reduced copy of the whole figure. In this Unit, three out of four quadrants on a square template were used to provide a basis for the parts that generate the self-similarity structure. The presence of a line symmetry in the whole figure implies that a given reduced copy within one of these three selected quadrants can be oriented by using (i) the identity mapping or a rotation, or by using (ii) a reflection. In short, two transformations exist for each of the three parts and either one of them could account for the orientation of the reduced copy within that part. Since one can select from two possible transformations in each of three parts, eight distinct geometric genetic codes exist that could generate the same symmetric fractal figure.

Changing the template for the construction process also implies that distinct geometric genetic codes can generate the same fractal image. Specifically, in Activities 8.4, 8.5, and 8.7 the top left, lower left, and lower right quadrants of a square template were used as a framework for iteration processes that led to self-similar fractals. If a different set of three quadrants are selected, then fractals images that appeared within the prior template configuration can recur in a different position on the square template. A reoriented version of a given fractal image might be described as a relative of the first figure. Not surprisingly, these reoriented versions are related to the original figure by one of the eight transformations of the square. Moreover, the geometric genetic codes of these relatives are algebraically related, a fact that will be explored in greater detail in Activity 8.9.

Implicit Discoveries. There are 8 symmetric and 448 non-symmetric fractal images that fit in the standard three quadrants of the template. This total gets multiplied by 4 when relatives are counted that contain images in the fourth quadrant.

8.9 Originals and Relatives

Specific Directions. In Activities 8.4, 8.5, and 8.7 the top left, lower left, and lower right quadrants of a square template were used as a framework for iteration processes that led to self-similar fractals. If a different set of three quadrants are selected, then fractals images that appeared within the prior template configuration recur in a different position on the square template. A reoriented version of a given fractal image might be described as a relative of the first figure. Activity 8.8 showed that these reoriented versions are related to the original figure by one of the eight transformations of the square. In this Activity, the geometric genetic codes of these relatives are shown to be algebraically related by a process called conjugation.

Suppose that an ORIGINAL fractal, such as the one shown in the illustration to the right, is reoriented to form a fractal IMAGE by transformation T where T is one of the eight motions of a square.

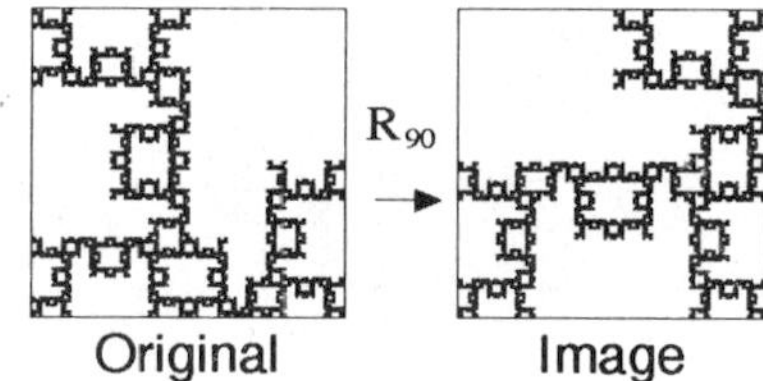

For example, the geometric genetic code of the illustrated ORIGINAL figure for quadrants a, b, and c, respectively, is (D^-, R_{90}, V). Then, under 90 degree rotation T, the quadrants a, b, and c in the ORIGIANL are rotated to quadrants b, c, and d in the IMAGE. The geometric genetic code of the IMAGE for quadrants b, c, and d, respectively, is (D^+, R_{90}, H). Once transformation T is identified, we can compute the geometric genetic code for the IMAGE figure directly by employing T and inverse mapping T^{-1}. Specifically, in the example, the inverse of R_{90} is R_{270}. The following compositions compute the IMAGE code directly from the code of the ORIGINAL:

$$R_{90}D^-R_{270} = D^+ \qquad R_{90}R_{90}R_{270} = R_{90} \qquad R_{90}VR_{270} = H$$

The IMAGE transformations D^+, R_{90}, and H are called conjugates of the ORIGINAL transformations D^-, R_{90}, and V, respectively, by R_{90}.

Implicit Discoveries. The Geometric Genetic Code of an original fractal figure and a relative (obtained via one of the eight transformations of the square) is algebraically related to the Geometric Genetic Code of the relative by the algebraic process of conjugation.

8.1 SYMMETRIES 8.1A

If a figure has line symmetry, then there are two halves of the figure that are images of each other when reflected about the line of symmetry.

If a figure has rotational point symmetry, then the figure can be turned back onto itself by rotating it less than one full turn about the point of symmetry.

1. This triangle can be rotated 120 or 240 degrees clockwise about its rotational point of symmetry and still have it appear in the position shown.

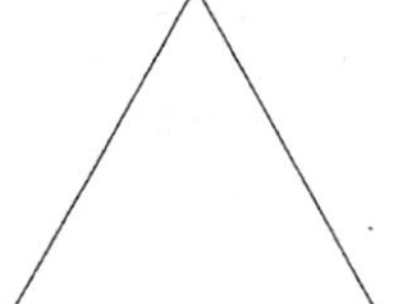

   a. Mark in the triangle its point of rotational symmetry.
   b. Draw through the triangle its three lines of symmetry.
   c. How are the lines of symmetry related to the rotational point of symmetry in an equilateral triangle?

2. Draw in all lines and rotational points of symmetry in each figure.

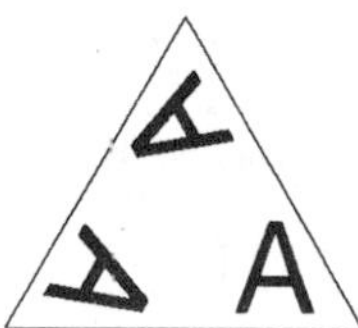

3. Some letters of the alphabet possess line symmetry, some rotational point symmetry, some both, and some neither.

   A B C D E F G H I J K L M N O P Q R S T U V W X Y Z

   a. Circle those letters that possess line symmetry.
   b. Underline those letters that possess rotational point symmetry.
   c. List all those letters that have neither line nor rotational point symmetry.

4. This square can be rotated clockwise by three different amounts less than one full turn and still have the square appear in the position shown.

   a. Mark in the square the point of rotational symmetry and give the number of degrees in each of these three possible turns.

   b. Draw through the square all lines of symmetry.

8.1B

The symmetries of a figure correspond to physical motions that map the figure back into itself. The IDENTITY motion I is a special null motion that maps the figure identically back into itself without any rotation about a point or reflection about a line.

A square has eight symmetries. These symmetries can be identified as the identity motion, counterclockwise rotations of 90, 180, and 270 degrees about the center point, and reflections about the horizontal and vertical axes and the two diagonals. We track these eight motions of the square here by using the small shaded corner rectangle.

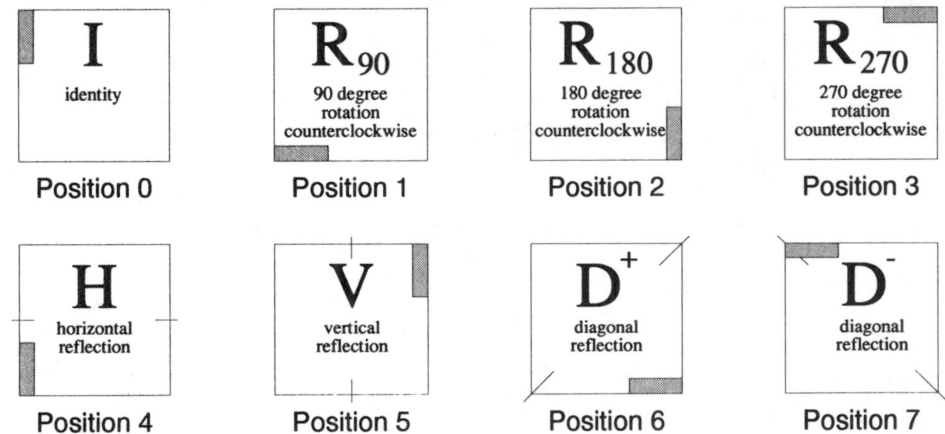

5. Draw in the letter **P** for the rotations and reflections indicated.

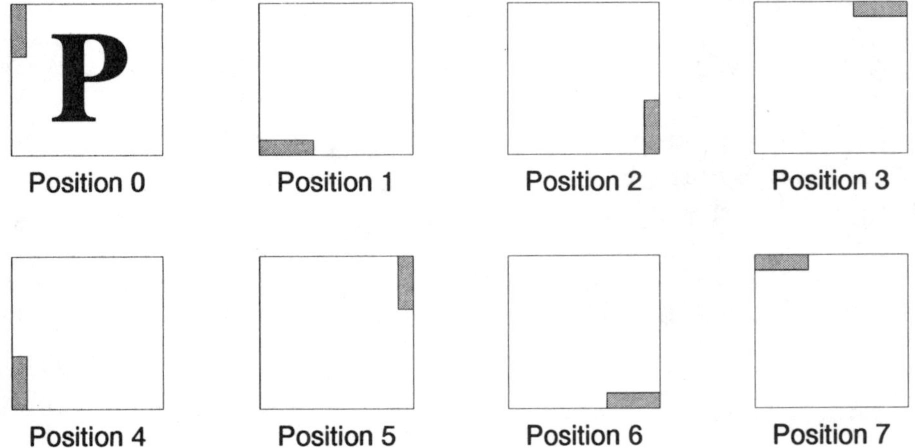

6. Let the figure in the first square represent the identity motion. Draw the figure in the remaining seven positions of the square, identifying the corresponding motions.

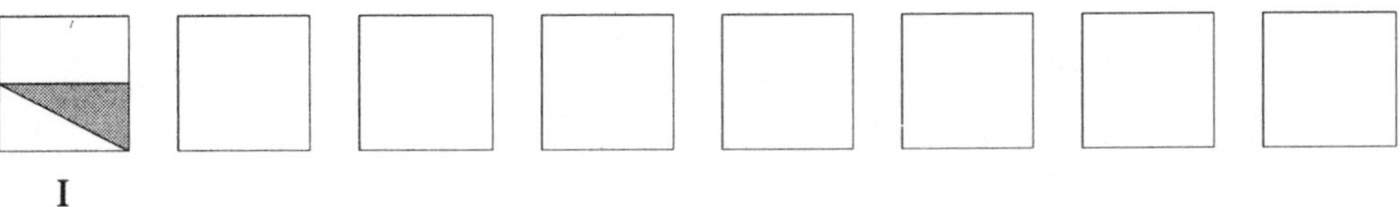

8.2 COMPOSITIONS

8.2A

A square may be mapped back into itself through turns and through flips. In all, there are eight possible motions that include rotations about the center point of symmetry, reflections about lines of symmetry, and the identity motion.

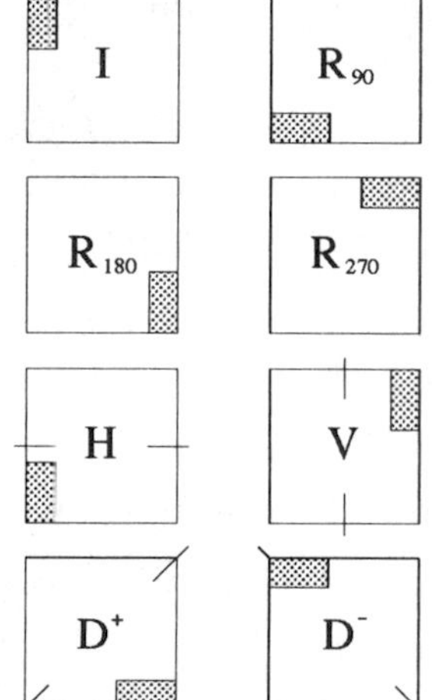

I	Identity motion
R_{90}	Counterclockwise rotation of 90°
R_{180}	Counterclockwise rotation of 180°
R_{270}	Counterclockwise rotation of 270°
H	Reflection about the horizontal axis
V	Reflection about the vertical axis
D^+	Reflection about the top-right bottom-left diagonal
D^-	Reflection about the top-left bottom-right diagonal

This activity addresses the question of composition of these symmetry transformations. What single motion is equivalent to a sequence of two or more successive motions?

Find the single symmetry transformation equivalent to

1. a 270° rotation followed by a 180° rotation
2. a 270° rotation followed by an H reflection about the horizontal axis
3. a D^+ reflection followed by a D^- reflection

Clearly, some compositions of motions are easier to see than others. Several methods can be used in finding the results. In every case, the composition can be expressed symbolically.

- *visualization* — through a mental image
- *manipulation* — through a hands-on model

We can visualize a 270° rotation followed by a 180° rotation resulting in the same position as a single 90° motion.

$$R_{180} \circ R_{270} = R_{180}[R_{270}] = R_{90}$$

8.2B

We can cut out and mark a paper square and then move it to show that successive steps of a 270° rotation followed by a reflection about the horizontal axis results in the same position as a single D^+ motion.

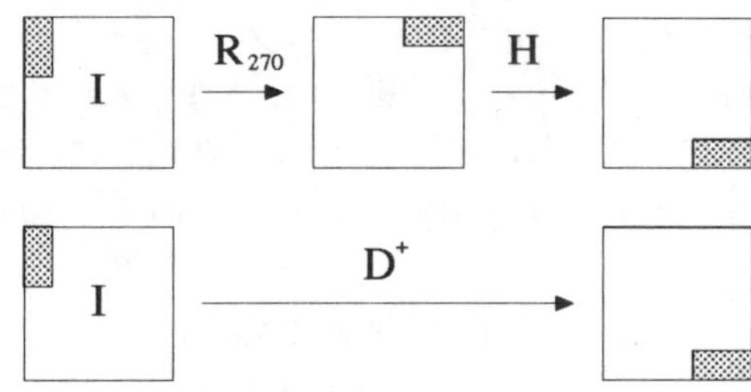

$$H \circ R_{270} = H[R_{270}] = D^+$$

Which method would you use to show that successive motions through D^+ and then D^- results in the same position as a single motion of R_{180}?

$$D^- \circ D^+ = D^-[D^+] = R_{180}$$

4. Is $H \circ R_{270}$ the same as $R_{270} \circ H$?
5. Is $D^-[D^+]$ the same as $D^-[D^+]$?

Any figure can be placed in the square being transformed. For example, consider the following example.

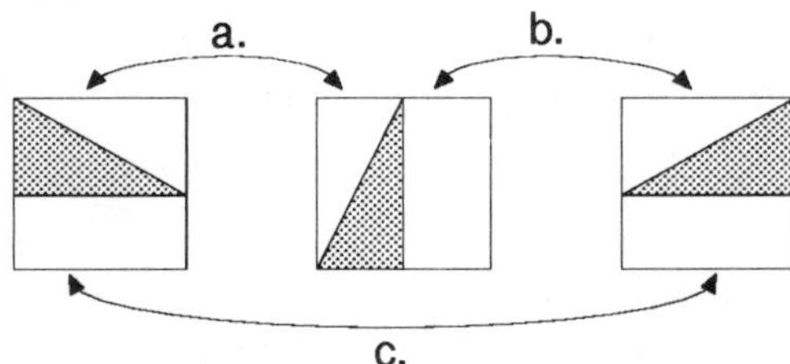

The transformations and the compositions are:
a. R_{90} b. D^- c. V Formula $D^- \circ R_{90} = D^-[R_{90}] = V$

Identify the motions and give the composition equation as in the example above.

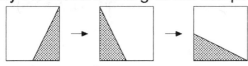

6. a. _____ b. _____ c. _____ Formula ____________________

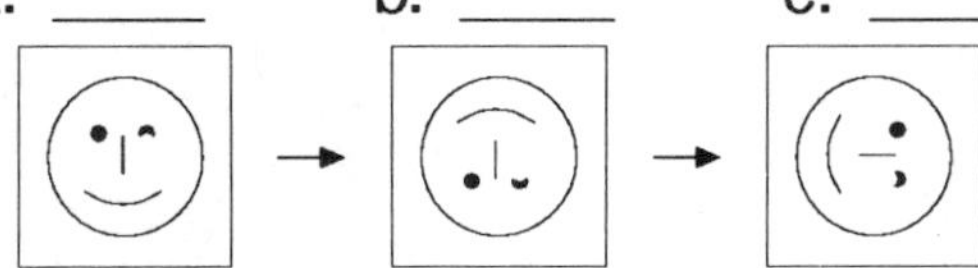

7. a. _____ b. _____ c. _____ Formula ____________________

8. a. _____ b. _____ c. _____ Formula ____________________

8.2C

The symmetry computer at the bottom of the page can be used as a hands-on aid to help when visualizing difficult compositions of motions. Make a cut-out copy of the shaded figure shown to the left below. Place this mask over the symmetry computer. Always start from the identity position I. Perform the sequence of rotation and reflection operations on the mask and read the results directly from the key .

9. Use the symmetry computer to find the resultant for each motion pair. Label the diagram accordingly.

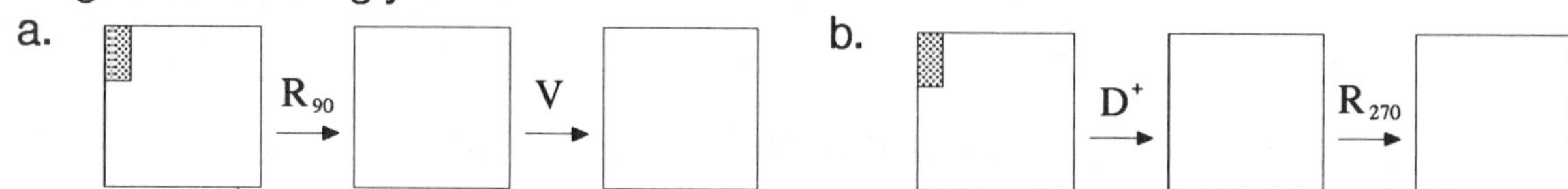

10. Find the single motion equivalent to each composition. Visualize first and then check using the symmetry computer.

a. $R_{180} \circ V = R_{180}[V] =$ _______

b. $D^{-} \circ R_{270} = D^{-}[R_{270}] =$ _______

c. $H \circ R_{90} = H[R_{90}] =$ _______

The initial position of the square is assumed to be the identity motion. On the symmetry computer, this has the rectangular cut-out running vertically in the upper left-hand position. The corresponding small shaded rectangles used in the squares in the art serve only to maintain orientation. Different initial positions of the square will yield different results for the same operation, as shown here.

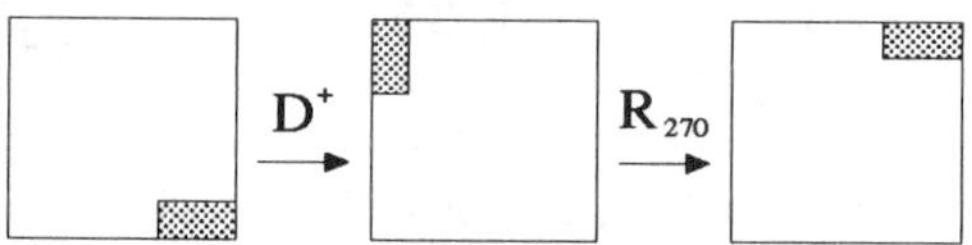

Manipulative: THE SYMMETRY COMPUTER

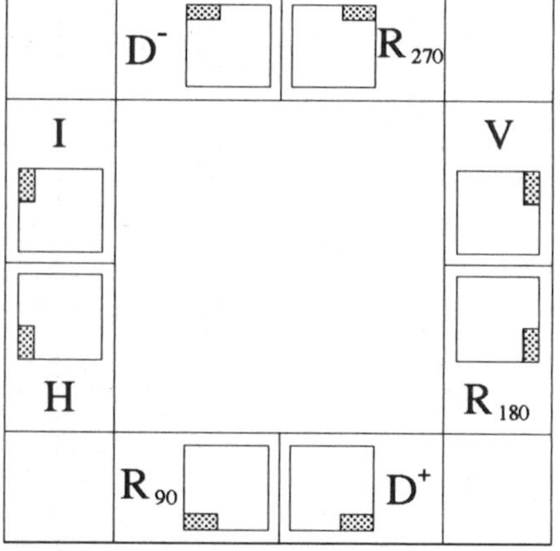

8.2D

Here are the eight possible positions of a figure in a square. Apply the given symmetry transformation or composition mapping to the specified initial figure (identity motion). Find the resulting figure. For example, transformation V applied to initial figure b gives c. Visualize first. Then use the symmetry computer as a check in the case of compositions.

a.  b. 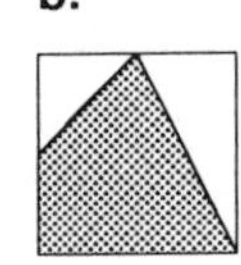 c. 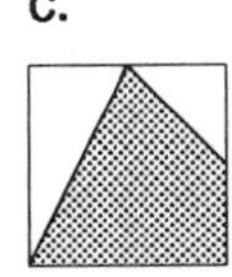 d. 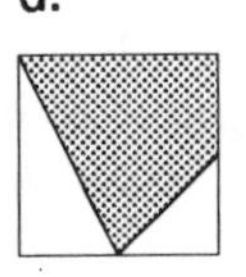 e. 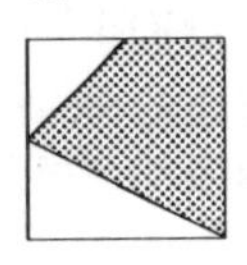 f. 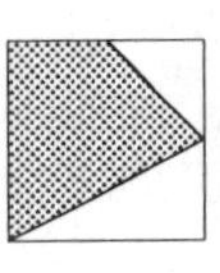 g. 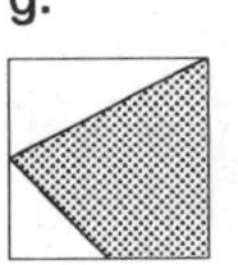 h. 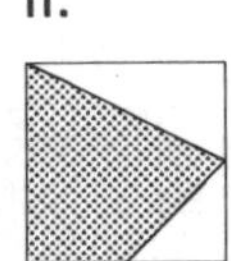

11. Use figure c as the initial figure and H as the transformation.

12. Use the initial figure g and D^+ as the transformation.

13. Apply the composition equation $R_{180} \circ V$ on the initial figure d.

14. Apply the composition equation $D^- \circ R_{270}$ on the initial figure b.

In dealing with symmetric transformations of a square, there are eight possible motions. Thus, in the composition of two motions, there are 8×8 or 64 possible combinations. Two results have already been entered in the appropriate cells of the table below. Recall that the notation used in the equation to indicate the order of composition follows the right-to-left convention.

Rotate 270° and then reflect about the horizontal axis: $H \circ R_{270} = D^+$
Reflect about vertical axis and then about top-left bottom-right diagonal: $D^- \circ V = R_{90}$
First the T_1 operation and then T_2: $T_2 \circ T_1$

15. Complete the table. Use the symmetry computer as needed.

$T_2 \circ T_1$		T_2 = second operation							
		I	R_{90}	R_{180}	R_{270}	H	V	D^+	D^-
	I								
	R_{90}								
T_1	R_{180}								
first	R_{270}					D^+			
oper-	H								
ation	V								R_{90}
	D^+								
	D^-								

8.3 ALGEBRAIC MAPPINGS

8.3A

A geometric transformation can be viewed as an algebraic mapping through the use of a coordinate system. The specific equations used to describe the mapping depend upon the placement of the axes. The eight symmetries of a square are developed algebraically in this activity by locating the axes on the square as shown at the right. Here the rotations R_{90}, R_{180}, and R_{270}, are about the origin, and the reflections, H and V, are about the x- and y-axes.

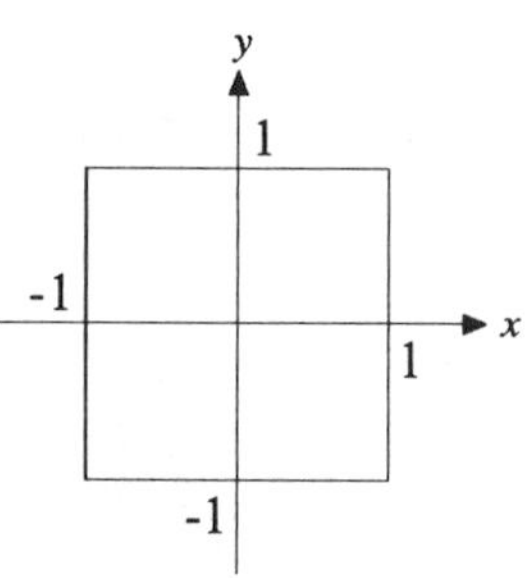

A geometric transformation maps every point (x, y) of the plane into a unique image point (x', y'). Note the new positions of the vertices of the shaded quadrilateral when the symmetry transformation H is applied to the square below.

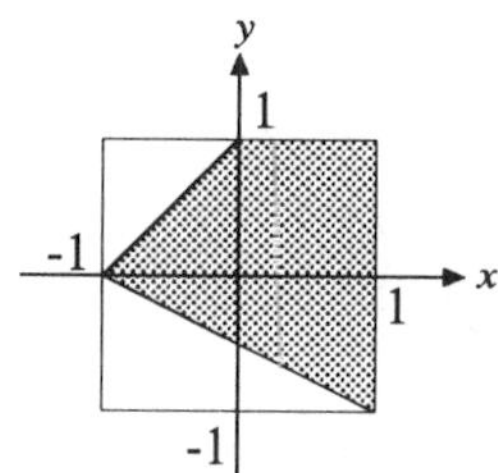

Symmetry Transformation

H

Reflect about the x-axis

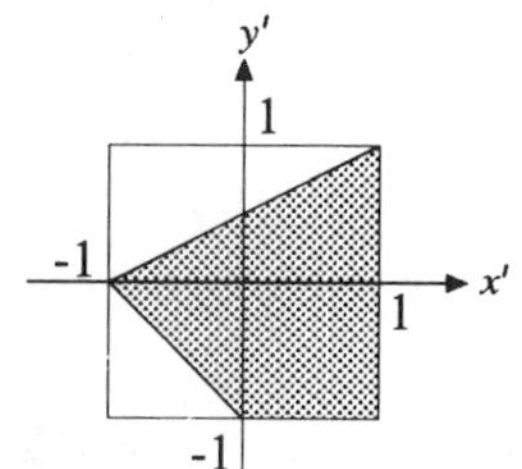

$(x, y) \to (x', y')$

$(-1, 0) \to (-1, 0)$
$(0, 1) \to (0, -1)$
$(1, 1) \to (1, -1)$
$(1, -1) \to (1, 1)$

1. Draw in the correct image for each symmetry transformation of the square and list the new coordinates for the vertices of the shaded quadrilateral.

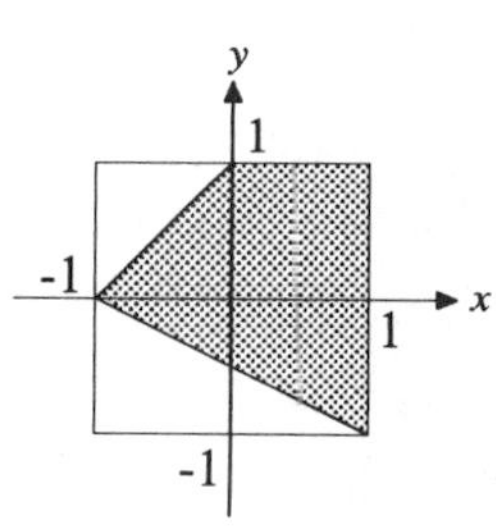

Symmetry Transformation

R_{270}

Rotate 270° about the origin

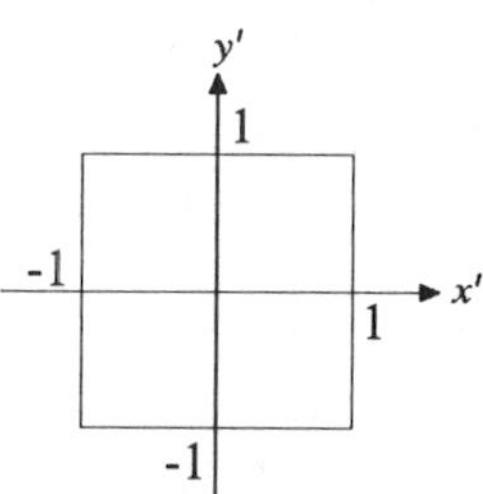

$(x, y) \to (x', y')$

$(-1, 0) \to (\_\_, \_\_)$
$(0, 1) \to (\_\_, \_\_)$
$(1, 1) \to (\_\_, \_\_)$
$(1, -1) \to (\_\_, \_\_)$

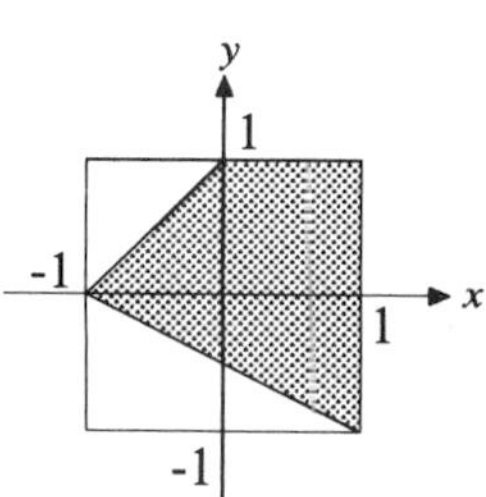

Symmetry Transformation

D^+

Reflect about the line $y = x$

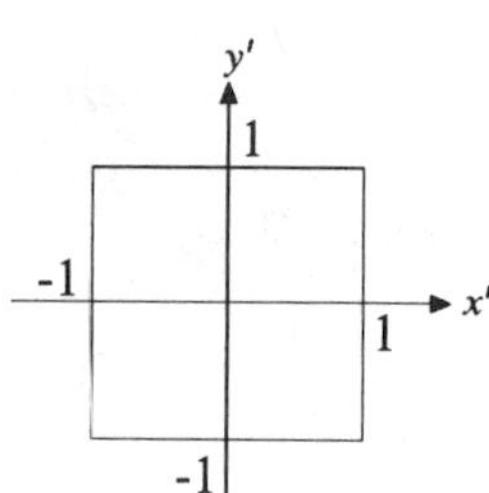

$(x, y) \to (x', y')$

$(-1, 0) \to (\_\_, \_\_)$
$(0, 1) \to (\_\_, \_\_)$
$(1, 1) \to (\_\_, \_\_)$
$(1, -1) \to (\_\_, \_\_)$

2. Find the image point (x', y') for each initial point (x, y).
   a. $(0, 0)$ under the rotation R_{180}
   b. $(0, -0.5)$ under the reflection V
   c. $(-0.5, 0.5)$ under the rotation R_{90}
   d. $(-0.5, -0.25)$ under the reflection D^+

8.3B

3. Draw in the correct image for the combined transformations of the square. List the new coordinates for the vertices of the shaded quadrilateral.

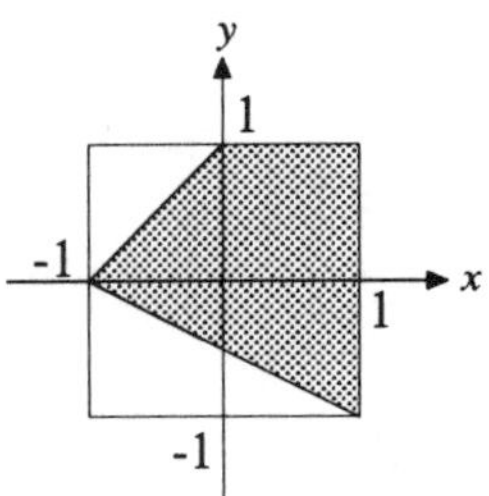

Symmetry Transformation

D^-

followed by

H

$(x, y) \to (x', y')$

$(-1, 0) \to (\_\_, \_\_)$
$(0, 1) \to (\_\_, \_\_)$
$(1, 1) \to (\_\_, \_\_)$

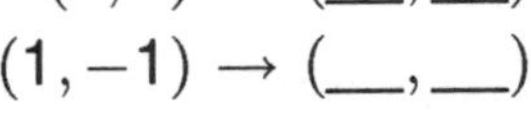
$(1, -1) \to (\_\_, \_\_)$

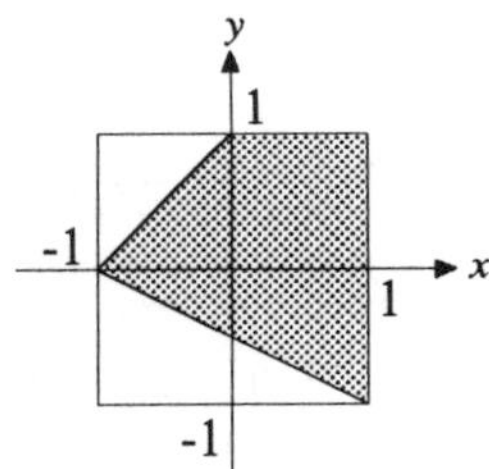

Symmetry Transformation

H

followed by

D^-

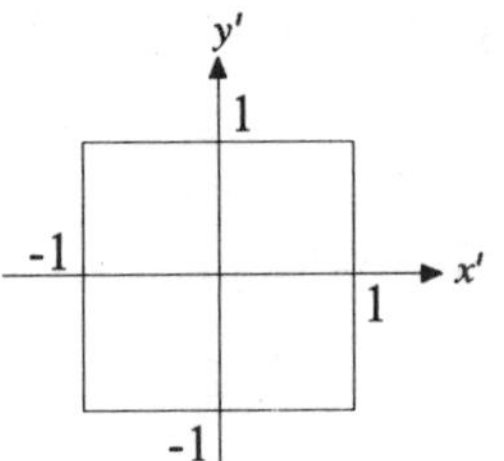

$(x, y) \to (x', y')$

$(-1, 0) \to (\_\_, \_\_)$
$(0, 1) \to (\_\_, \_\_)$
$(1, 1) \to (\_\_, \_\_)$
$(1, -1) \to (\_\_, \_\_)$

An algebraic representation of a mapping can be expressed in terms of a system of linear equations or the corresponding matrix equation form.

$$x' = px + qy \qquad \begin{pmatrix} x' \\ y' \end{pmatrix} = \begin{pmatrix} p & q \\ r & s \end{pmatrix} \begin{pmatrix} x \\ y \end{pmatrix}$$
$$y' = rx + sy$$

4. Compare the two algebraic forms of the same general transformation shown above. How can you quickly translate from the matrix to the corresponding linear form?

Specific values of the parameters p, q, r, and s produce specific transformations. Of special interest here are those linear and matrix equations that belong to the eight symmetries that map the square back into itself. What are the required values of p, q, r, and s for these important transformations?

5. Use the linear equations $x' = 1x + 0y$ and $y' = 0x - 1y$ to find the image point (x', y') for each initial point (x, y). What symmetry transformation do these equations appear to produce?

a. $(1, 1)$ b. $(-1, 1)$ c. $(1, -1)$ d. $(-1, -1)$

6. Use the matrix equation $\begin{pmatrix} x' \\ y' \end{pmatrix} = \begin{pmatrix} 0 & 1 \\ -1 & 0 \end{pmatrix} \begin{pmatrix} x \\ y \end{pmatrix}$ to find the image point (x', y') for each initial point (x, y). What symmetry transformation do these equations appear to produce?

a. $(1, 1)$ b. $(-1, 1)$ c. $(1, -1)$ d. $(-1, -1)$

8.3C

7. Write the identity matrix that maps each point of the square back into itself.

8. Do the linear equations $x' = 0x + 1y$ and $y' = 1x + 0y$ have the effect of interchanging the x and y coordinate values of each point? This is equivalent to reflecting the square about what line? What symmetry mapping of the square must it represent?

9. Show that the matrix $\begin{pmatrix} 0 & -1 \\ -1 & 0 \end{pmatrix}$ reflects the four vertices of the square about the line $y = -x$, and thus appears to be the matrix associated with the D^- symmetry of the square.

10. A 90° counterclockwise rotation maps the point (x, y) into $(-y, x)$. Show this mapping as a matrix equation.

11. What matrix gives the R_{180} mapping of a square?

12. What matrix gives the V mapping of a square?

There is a unique matrix associated with the mapping for each of the eight symmetries of the square. Both the geometric and algebraic mappings are listed here.

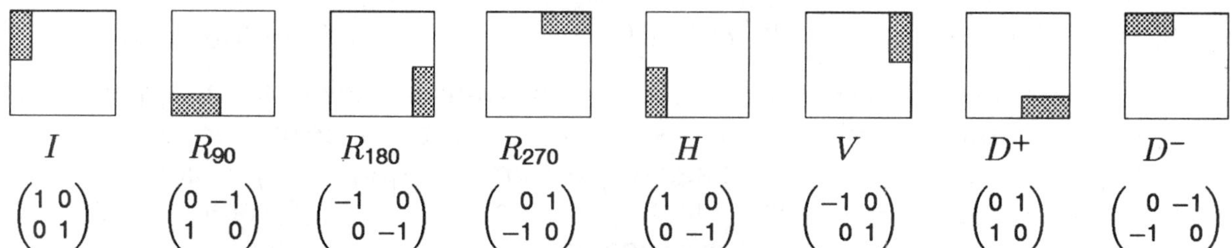

13. Use a matrix to perform the reflection mapping V on the point $(-1, 1)$.

$$\begin{pmatrix} \_\_ & \_\_ \\ \_\_ & \_\_ \end{pmatrix} \begin{pmatrix} -1 \\ 1 \end{pmatrix} = \begin{pmatrix} \_\_ \\ \_\_ \end{pmatrix} \qquad V(-1, 1) = (\_\_, \_\_)$$

14. Use a matrix to perform the rotation mapping R_{270} on the point $(-1, 1)$.

$$\begin{pmatrix} \_\_ & \_\_ \\ \_\_ & \_\_ \end{pmatrix} \begin{pmatrix} -1 \\ 1 \end{pmatrix} = \begin{pmatrix} \_\_ \\ \_\_ \end{pmatrix} \qquad R_{270}(-1, 1) = (\_\_, \_\_)$$

The composition $R_{270} \circ V(x, y)$ means that the rotation R_{270} is applied to (x', y'), the result of the reflection V, producing (x'', y'').

$$R_{270} \circ V(x, y) = (x'', y'')$$

$$(x', y') = V(x, y)$$
$$(x'', y'') = R_{270}(x', y') = R_{270}(V(x, y))$$

15. Use the matrices for the composition given above to find the image point (x'', y'') if the initial point $(x, y) = (-1, 1)$. What single matrix can be used to find the combined effect of R_{270} applied to the result of V? Is it the same matrix that gives the combined effect of V applied to the result of R_{270}?

8.3D

16. Explain how the two matrices for R_{270} and V can be combined to find the single matrix D^{+} asked for in question 15.

$$R_{270} \circ V(-1,1) = R_{270}[V(-1,1)] = D^{+}(-1,1)$$

$$\begin{pmatrix} x'' \\ y'' \end{pmatrix} = \begin{pmatrix} 0 & 1 \\ -1 & 0 \end{pmatrix} \left\{ \begin{pmatrix} -1 & 0 \\ 0 & 1 \end{pmatrix} \begin{pmatrix} x \\ y \end{pmatrix} \right\} = \begin{pmatrix} 0 & 1 \\ -1 & 0 \end{pmatrix} \begin{pmatrix} -x \\ y \end{pmatrix} = \begin{pmatrix} y \\ x \end{pmatrix} = \begin{pmatrix} 0 & 1 \\ 1 & 0 \end{pmatrix} \begin{pmatrix} x \\ y \end{pmatrix}$$

Use the same method to find the single equivalent matrix for $V \circ R_{270}$.

The symmetry computer at the bottom of the page can be used as a hands-on aid to find the single equivalent transformation and its matrix for any composition sequence. Make a cut-out copy of the shaded figure shown to the left below. Place this mask over the symmetry computer. Always start from the identity position I. Perform the sequence of rotation and reflection operations on the mask and read the results directly from the key.

17. Use the symmetry computer to identify the single matrix that could be used to find the combined effect of the following compositions on the point (x, y). Verify each result algebraically.

a. $R_{90} \circ H(x,y)$ b. $D^{-} \circ V(x,y)$ c. $D^{+} \circ R_{180}(x,y)$

This activity shows how powerful the tools of mathematics can be. We began the unit showing eight symmetry transformations of a square geometrically. Here we find that through coordinate geometry, these same transformations and compositions of these motions can be treated algebraically through matrices. It is this algebraic approach that will be used when these mappings are executed through technology.

Manipulative: THE SYMMETRY COMPUTER

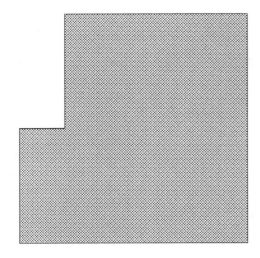

	$D^{-} \begin{pmatrix} 0 & -1 \\ -1 & 0 \end{pmatrix}$	$\begin{pmatrix} 0 & 1 \\ -1 & 0 \end{pmatrix} R_{270}$	
I $\begin{pmatrix} 1 & 0 \\ 0 & 1 \end{pmatrix}$			V $\begin{pmatrix} -1 & 0 \\ 0 & 1 \end{pmatrix}$
$\begin{pmatrix} 1 & 0 \\ 0 & -1 \end{pmatrix}$ H			$\begin{pmatrix} -1 & 0 \\ 0 & -1 \end{pmatrix}$ R_{180}
	$R_{90} \begin{pmatrix} 0 & -1 \\ 1 & 0 \end{pmatrix}$	$\begin{pmatrix} 0 & 1 \\ 1 & 0 \end{pmatrix} D^{+}$	

8.4 FRACTAL IMAGES

8.4A

Objects that are called fractal typically possess the property of self-similarity. When an object is self-similar, it can be partitioned into parts that together comprise the whole object in such a way that each of these parts is a replica of the whole object. In order to observe the self-similarity properties of these images, we need to uncover the hidden symmetries.

Study the image below on the left. Notice how the square cells lettered A, B, and C contain reduced replicas of the entire large square image transformed through some of the eight symmetries. In particular, cell C isolated at the right is a copy of the original square image reduced to half-size and transformed through a 180° rotation. Can you find which transformations have been used to produce the reduced copies in cells A and B?

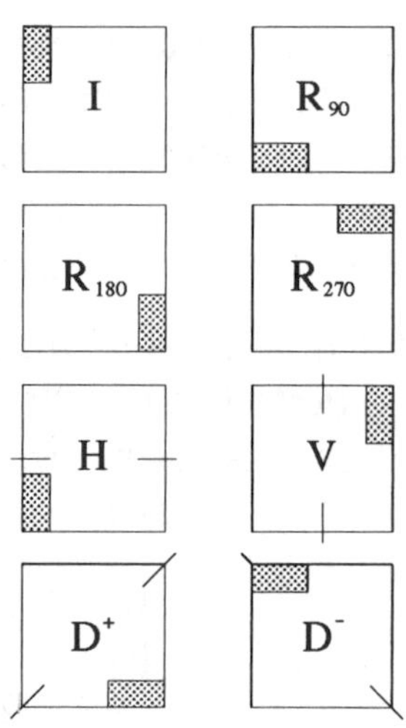

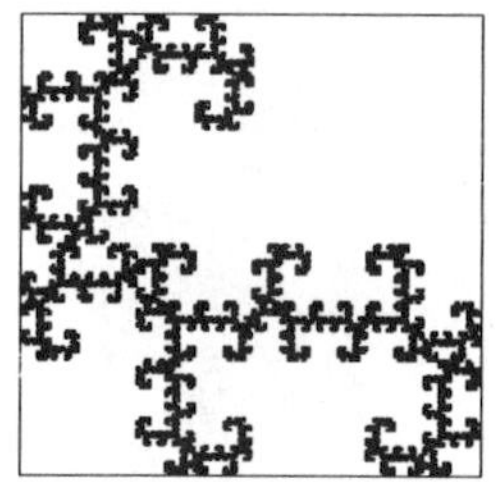

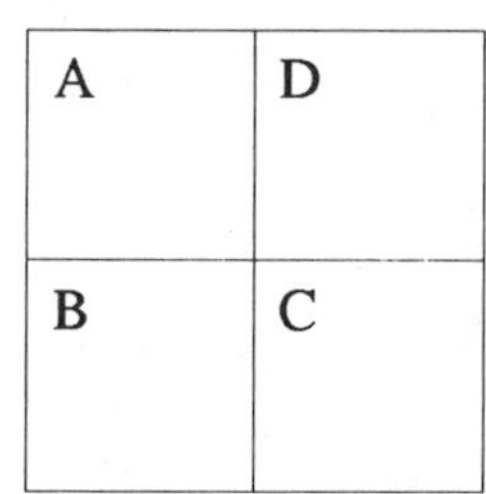

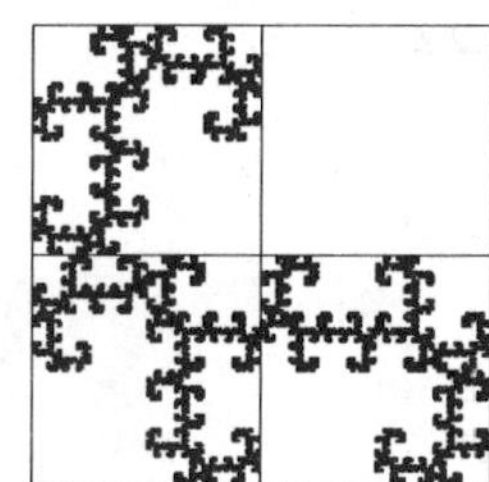

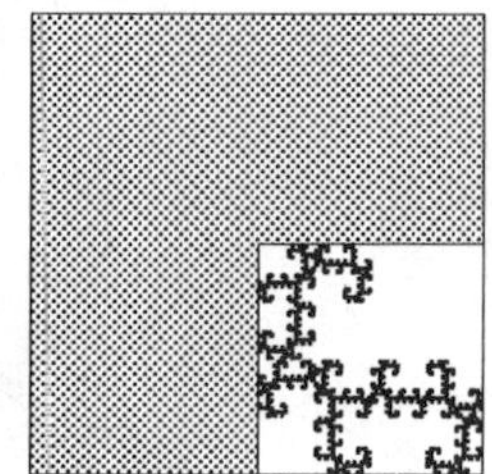

In going from one stage to the next in the iteration process that generated this fractal, copies were reduced, replicated, and rebuilt according to a specific algorithm defined by these three transformations. Each new stage became more delicate and intricate than the preceding one. The fractal is the final state. From that point on, repeated iterations produce exact, identical images possessing this property of self-similarity.

Here are two more examples illustrating, through symmetry transformations, the self-similarity of fractal images of this type.

Fractal Image | Subdivision | Symmetry Transformations

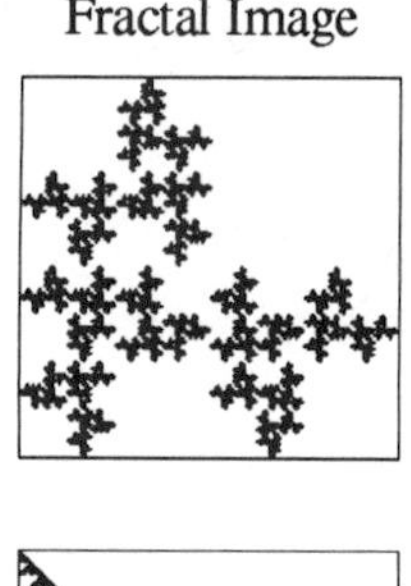

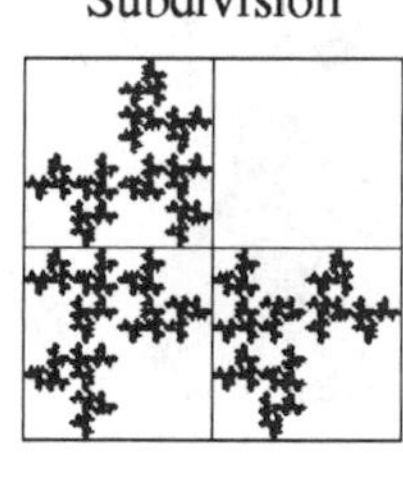

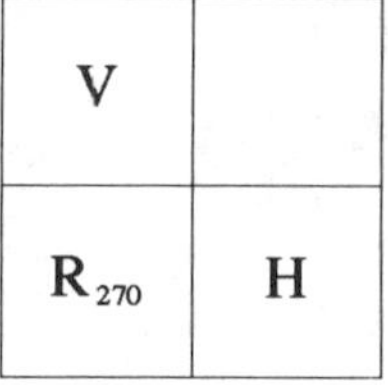

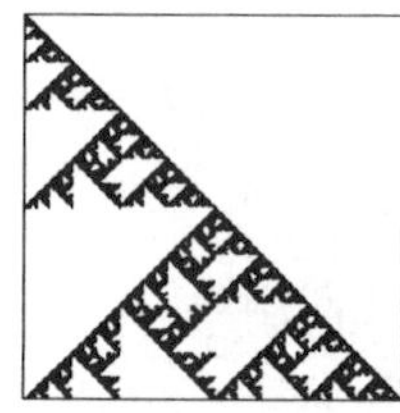

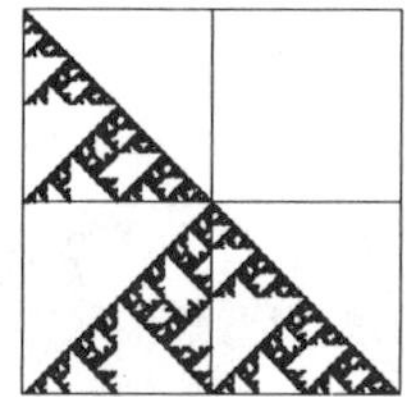

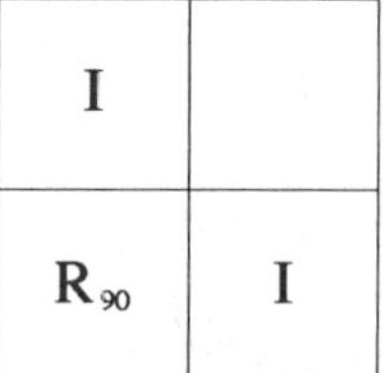

8.4B

Study the fractal images shown on the left below. Use the subdivision into the three smaller square cells to help uncover the hidden symmetries that identify the transformations used in each construction. Enter them, as shown, in the grids on the right. Think in terms of the small cells being reduced, replicated, and transformed versions of the original fractal image.

You may want to cut out and use the aids at the bottom of the page. Fold each one on the dotted line so that the reduced initial image is on one side and its reflection on the other side. See how they must be moved to match the images in cells A, B, and C.

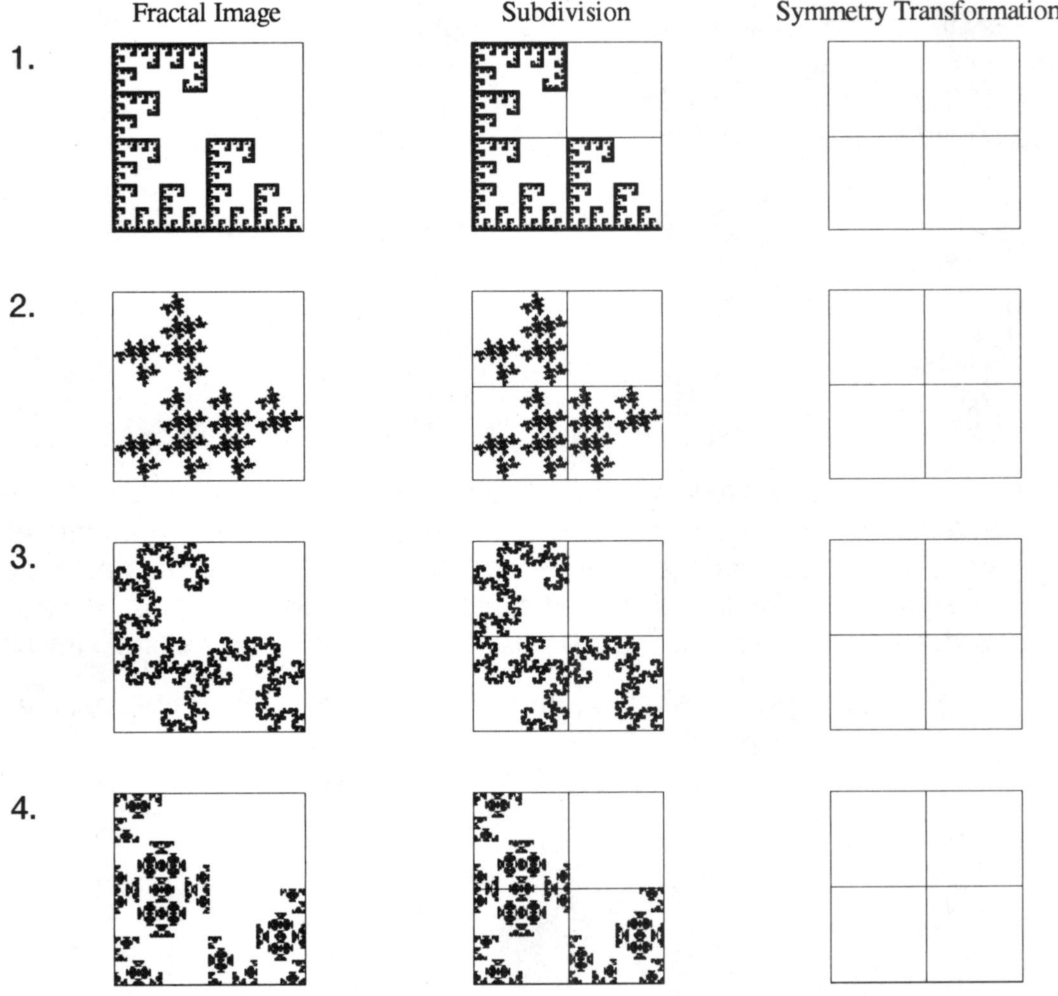

Manipulative:

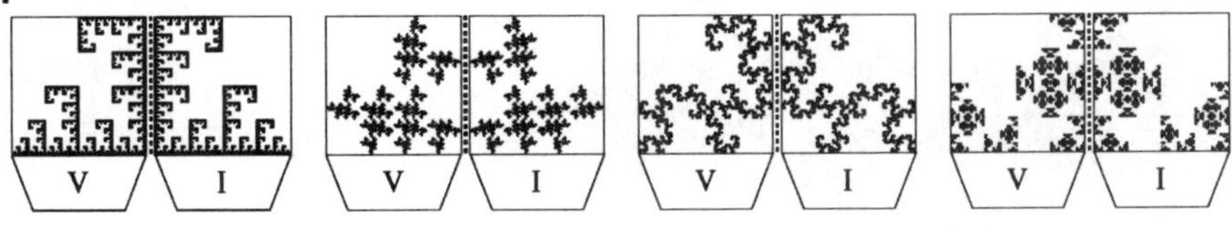

8.4C

This visual analysis suggests that we might describe each of the foregoing images by listing in order the three motions used to encode the image.

For example, the large image in the square to the left is composed by mapping three small copies of itself into cells A, B, and C using the three motions, V, R_{270}, and D^{-}, respectively. We can code this description using an ordered triple of symmetry transformation symbols. This is called the GEOMETRIC GENETIC CODE of the image. In one sense, this code characterizes the self-similarity features of the entire image.

Fractal Image	Subdivision	Symmetry Transformations	Geometric Genetic Code
	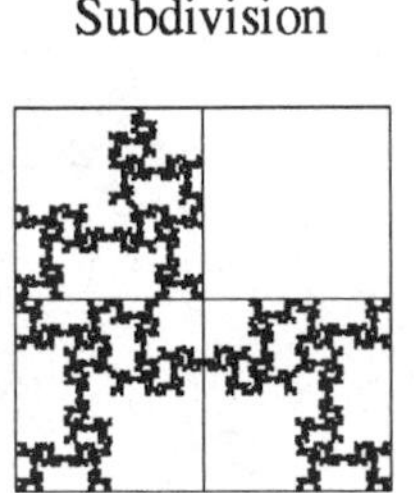	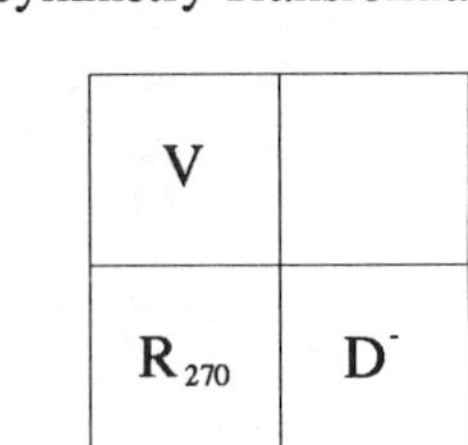	(V, R_{270}, D^{-})

5. Find the four geometric genetic codes for the fractal images in questions 1–4.

__________ __________ __________ __________

Find the geometric genetic codes used to build these fractal images. Subdivide each image into four smaller squares and view each cell as a symmetry transformation of a reduced version of the whole. Use the aids at the bottom, if needed.

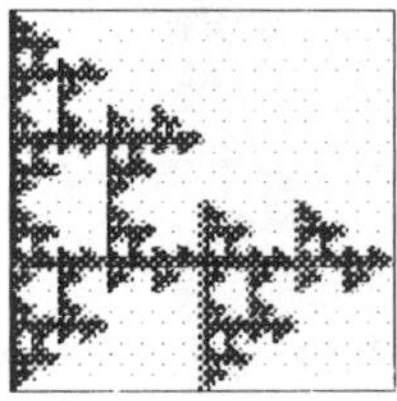
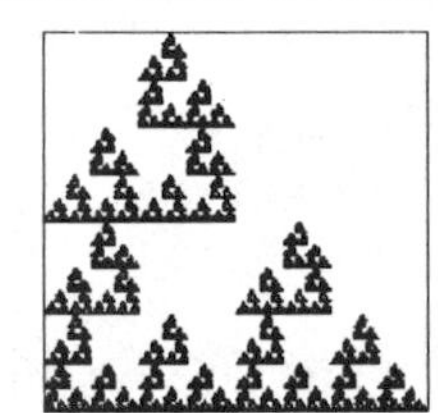

6. __________ 7. __________ 8. __________ 9. __________

10. Compare the initial image in question 6 with that in question 7. Which of the eight motions of the square would map image 6 into the position of image 7?

11. Compare the initial image in question 8 with that in question 9. Which of the eight motions of the square would map image 8 into the position of image 9?

Manipulative:

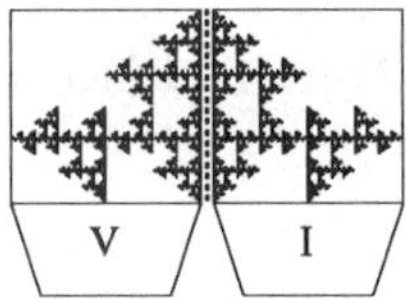

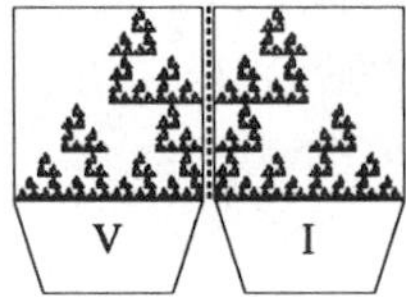

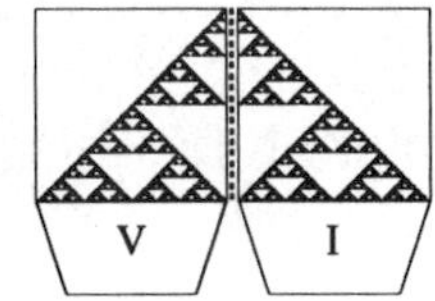

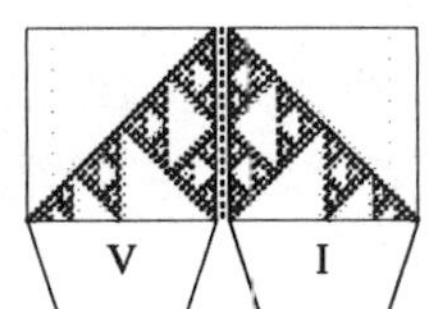

8.4D

Cell A **H**	
Cell B **?**	Cell C **H**

12. Simple changes in the geometric genetic code for a fractal can alter its appearance dramatically. For example, the fractals shown below all utilize the reflection transformation H in cells A and C. However, in each case a different one of the eight symmetry transformations of the square has been used in cell B. Find the geometric genetic code for each of these eight fractals.

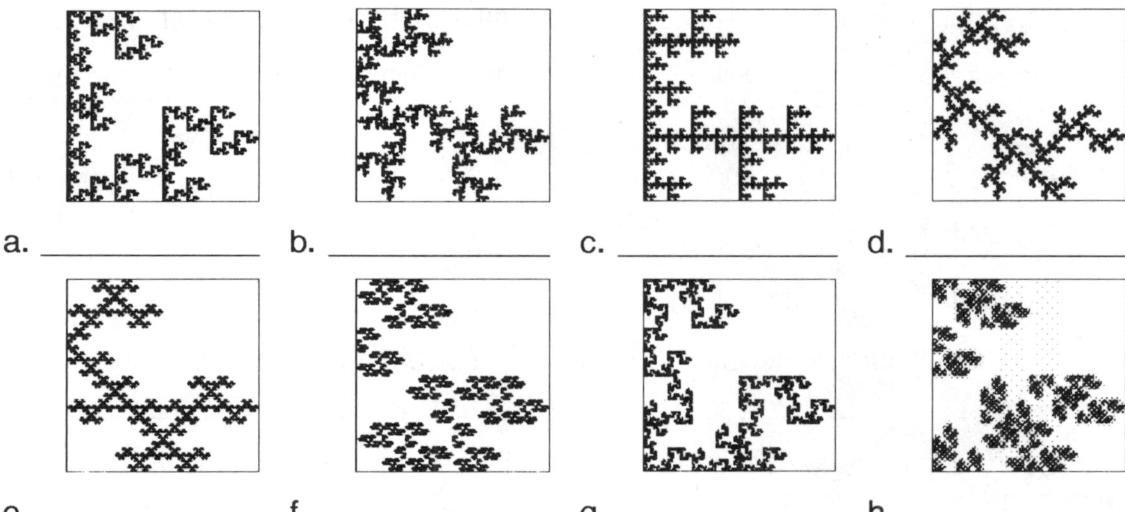

13. Each of the eight fractals shown above are different. Yet some have common characteristics that others do not have. Some are connected, others not, some contain "loops" and others do not. Write briefly on their similarities and on their differences.

14. Changing the order within the genetic code can also produce significant differences in the appearance of the fractal. The codes for these fractals each utilize the rotation transformation R_{270} twice and the reflection transformation V once. Which of the three possible codes belongs to each fractal?

a.

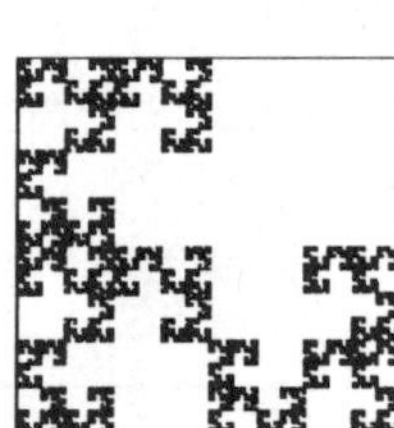

b.

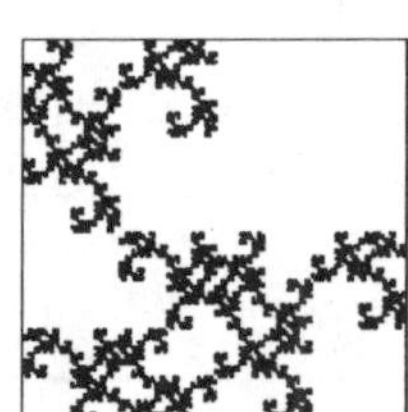

c.

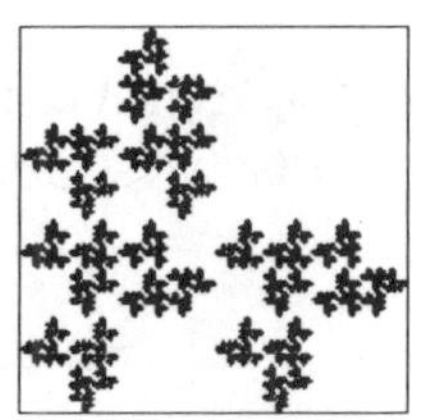

15. In some cases, two different geometric genetic codes produce the identical fractal. Think carefully about the specific properties of the mappings and see if you can suggest two codes that might have this property.

8.5 BUILDING FRACTALS

8.5A

One view of the iteration process for generating a fractal is to think of an endlessly repeating sequence where reduced copies are rebuilt by some well defined algorithm. In the illustration below, the original square with the letter P is reduced to half size. Three copies are made and rebuilt into a new square. The rebuilding into stage 1 is done by orienting each of the three reduced images by one of the eight possible motions of the square. The same process is then applied to the entire stage-1 figure to produce stage 2, to stage 2 to produce stage 3, and so on.

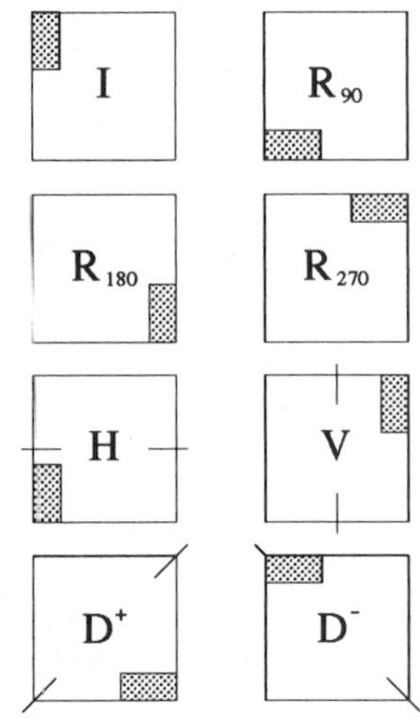

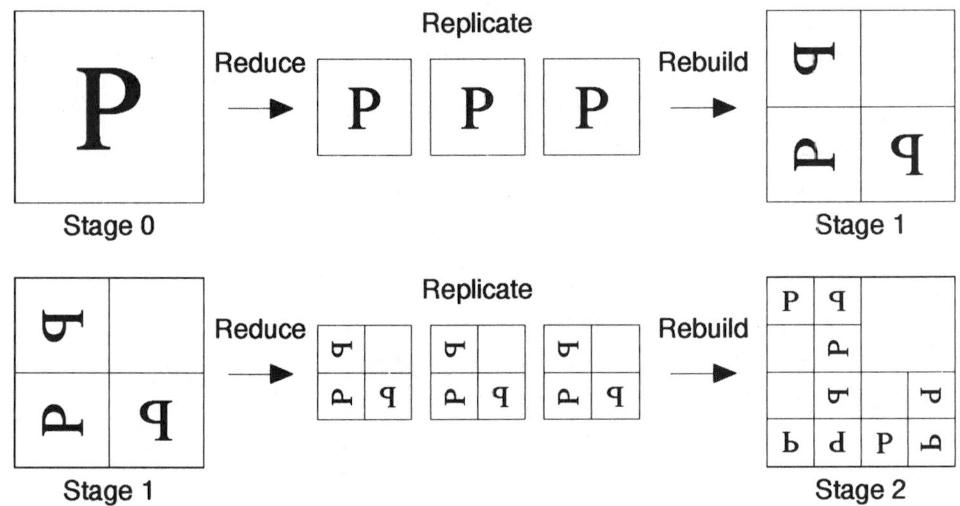

1. What symmetry transformation is used on each of the three reduced images in rebuilding from one stage to the next?
   Cell A ____ Cell B ____ Cell C ____

A	
B	C

2. Express this rebuilding algorithm using the notation of a geometric genetic code. Give each transformation as a letter triple, listing the motion of cell A first, cell B second, and cell C third.
   (____, ____, ____)

3. Apply the same building code a third time and construct stage 3.

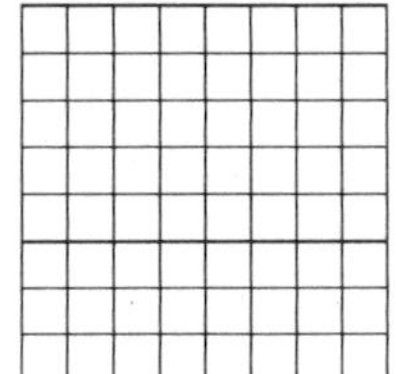

8.5B

The geometric genetic code is the building code. Applied over and over, always on the final image from the preceding stage, an underlying fractal image emerges.

4. One P appears at stage 0, three at stage 1, and nine at stage 2. How many are needed at stage 3? at stage 10? at stage n?

5. Are the positions of the letters important at every stage? Are the letters themselves important at every stage? Explain your answers.

The rebuilding step in this fractal generating process assigns a particular one of the eight motions of the square to each of the three cells, A, B, and C.

6. Suppose you could only select from the two symmetry transformations that reflect about the horizontal and vertical axes. Then only H or V could be assigned as a motion in cells A, B, and C. Complete the eight different choices that are possible by entering H or V into the appropriate cells.

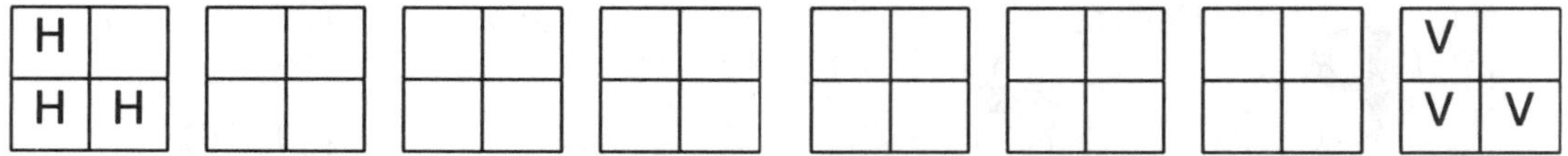

7. How many rebuilding choices are possible using only I, R_{90}, R_{180}, and R_{270} as motions for the cells A, B, and C?

8. This tree diagram lists all $2^3 = 8$ possible assignments of the motions H and V to the three cells. How many rebuilding choices are possible if any one of the eight symmetry transformations of the square are used as motions for the three cells? Note that this means there are only two choices for each of the three cells.

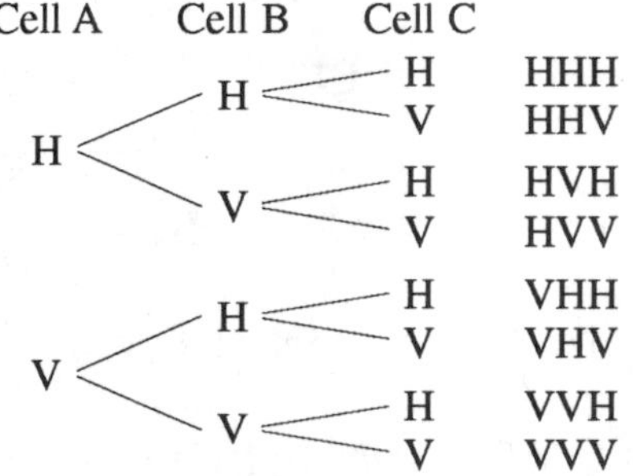

9. We select (H, V, V) one of the eight sets of symmetry transformations in the list that uses both H and V. Use it as the geometric genetic code and apply it to generate stages 2, 3, and 4 below, shading in entire pixels at each stage. Do you have a vision of the resulting fractal from the stage 4 figure?

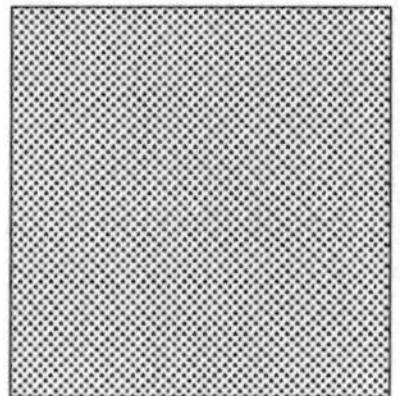
Stage 0

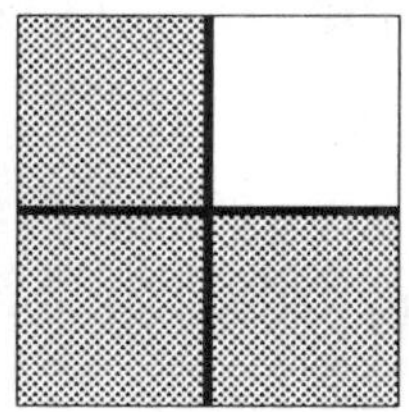
Stage 1

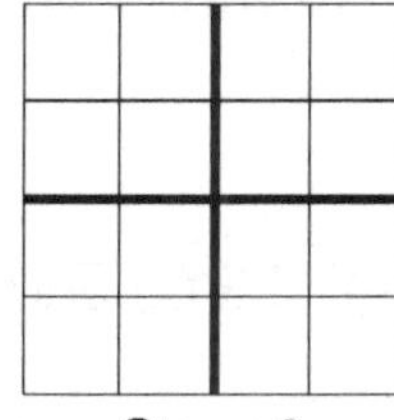
Stage 2

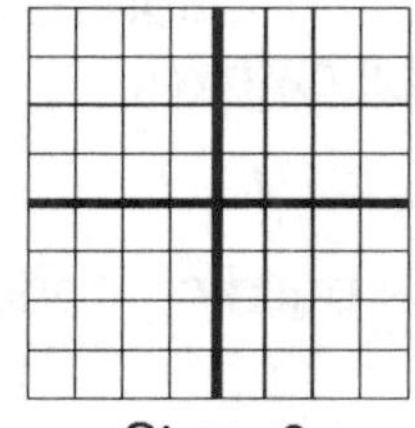
Stage 3

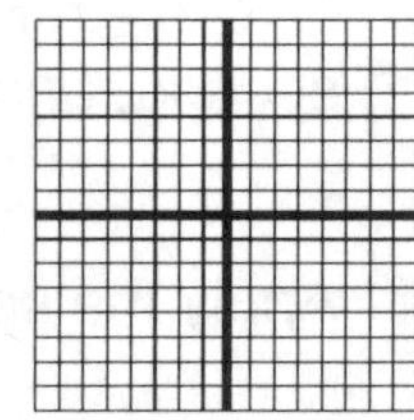
Stage 4

8.5C

10. Here are fractal images of all eight of the geometric genetic codes that use only the reflection transformations of H or V. Which one matches the completed stage-4 image created in question 9? Can you unravel the codes for the others?

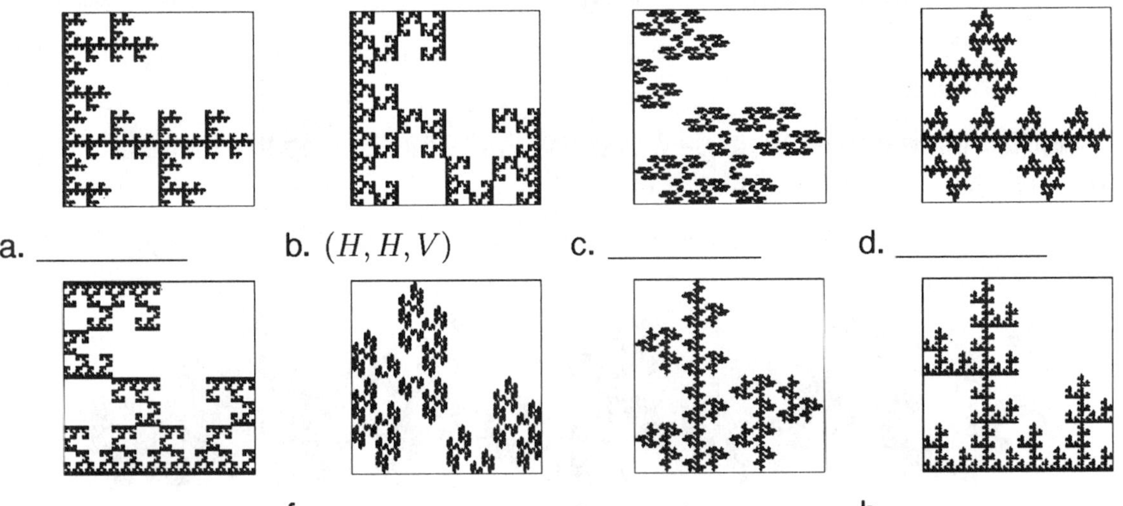

a. ________ b. (H, H, V) c. ________ d. ________

e. ________ f. ________ g. ________ h. ________

Some fractals have a built in symmetry that allows them to be generated by several different building codes. The Sierpinski triangle is one such example.

11. The first four stages shown below are clearly leading to the Sierpinski triangle. One possible geometric genetic code is (I, I, I). Another is (D^+, D^+, D^+). There are six more possibilities. Can you find them?

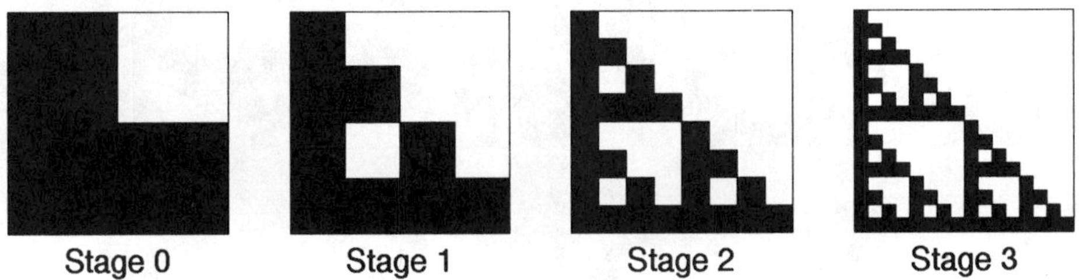

Stage 0 Stage 1 Stage 2 Stage 3

One way to help gain greater detail in generating a fractal image is to let it grow in size after a certain stage. This final activity generates the first four stages as before. But then it continues by replicating and rebuilding without reducing. Now the image exhibits greater and greater detail by growing in size.

12. Apply the building code (I, R_{180}, R_{90}) to generate stages 2, 3, and 4.

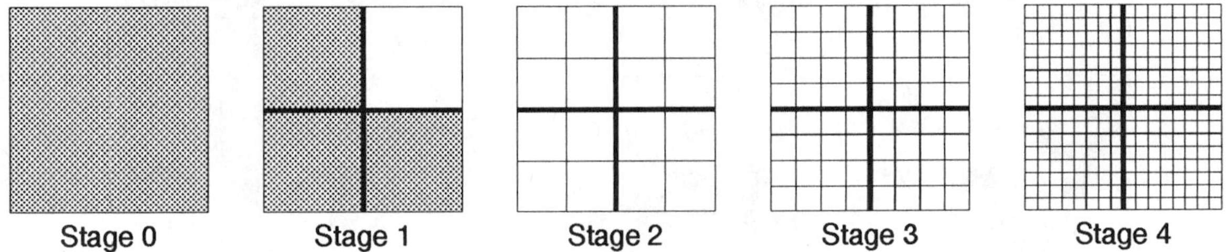

Stage 0 Stage 1 Stage 2 Stage 3 Stage 4

8.5D

Images for stage 4 of the building code (I, R_{180}, R_{90}) are reproduced below.

13. The image in a single square is stage 4. Cut out three of these squares, transform them appropriately, and tape them together to show stage 5, without reduction.
14. Construct three copies of stage 5 and correctly assemble them to show stage 6, again without reduction.
15. Construct three copies of stage 6 and assemble them using the same building code to construct stage 7, without reduction.
16. How many of the stage-4 squares would be needed to construct stage 10 for the eye to see?

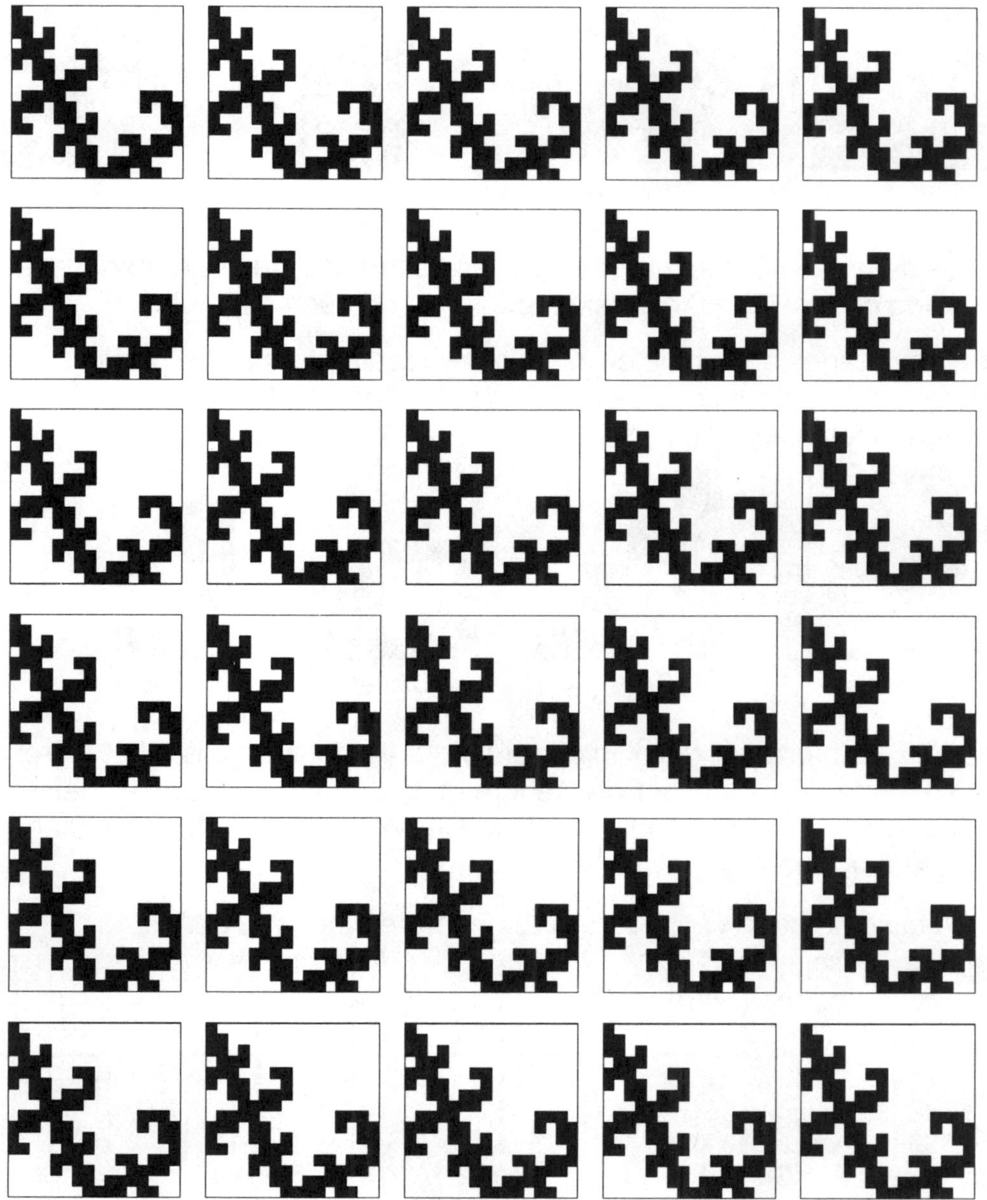

8.5E

In its final state, a fractal of this type contains three parts that are identical copies of the whole. This is the property of self-similarity.

17. Cut three copies from the set below and assemble them into a single, larger, similar structure. Repeat the process two additional times. Then combine these three larger parts into a still larger, similar structure of nine pieces. Keep combining parts, three at a time, with others through three more stages of growth.

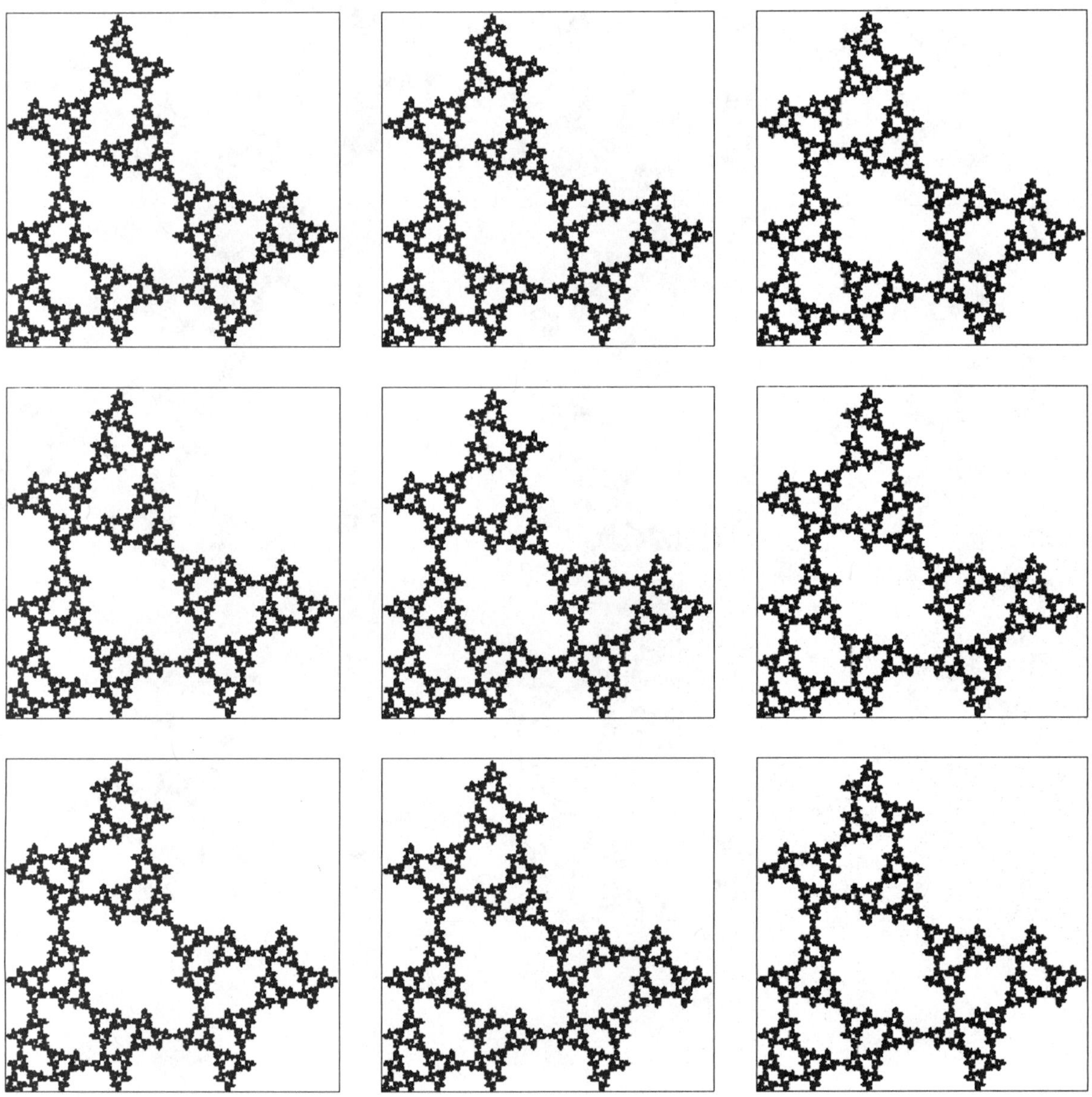

8.5F

18. Join with others to combine $3^5 = 243$ of the pieces below into a single fractal structure.

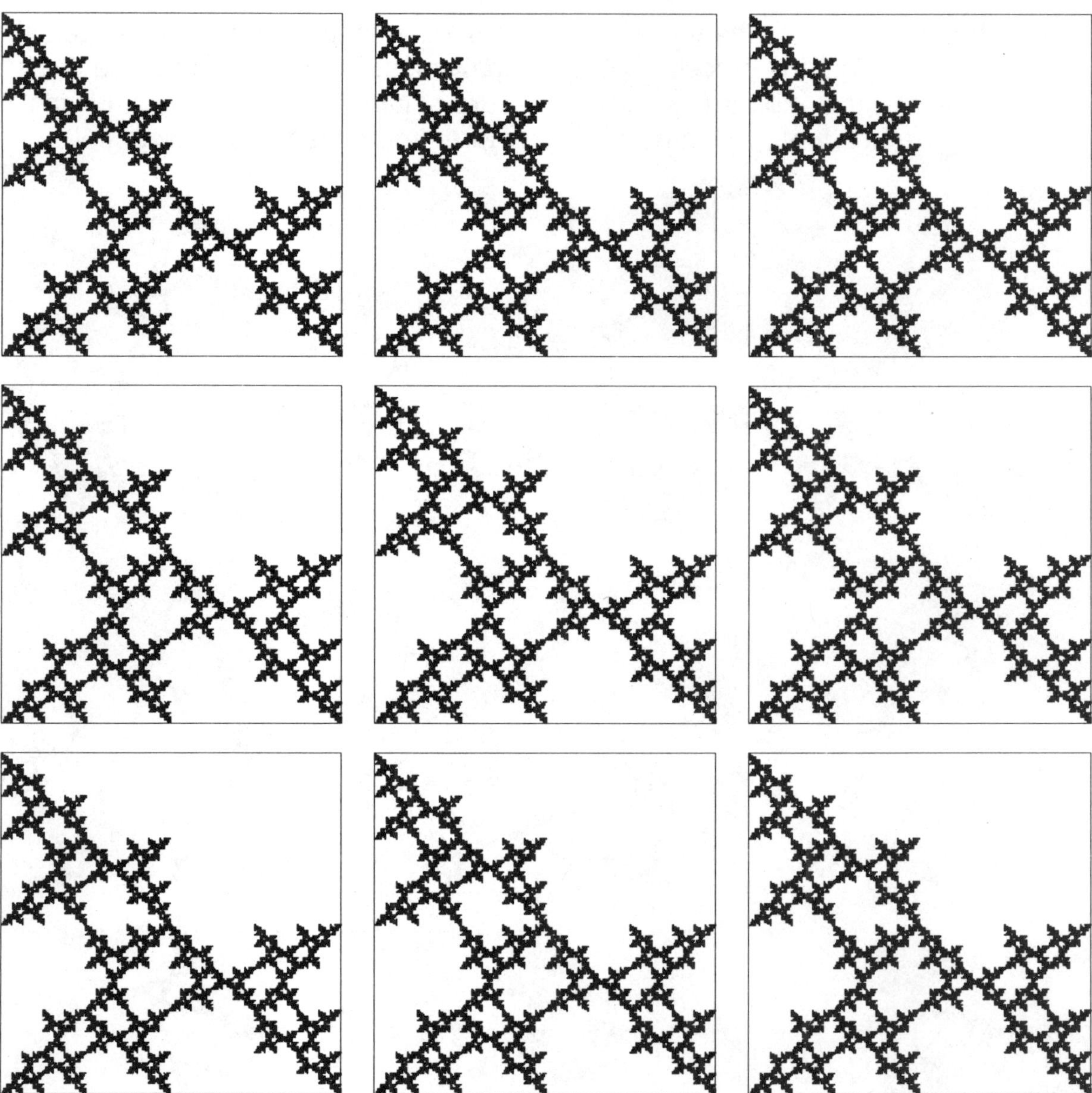

8.6 SUMMARIZING THE PROCESS 8.6A

The geometric structures of the two fractals shown below are quite different, as seen by their geometric genetic codes. However, the underlying processes that generated them are very much alike. By looking at the whole mechanism of the two-dimensional iterated function systems that were used, we will be able to see these similarities.

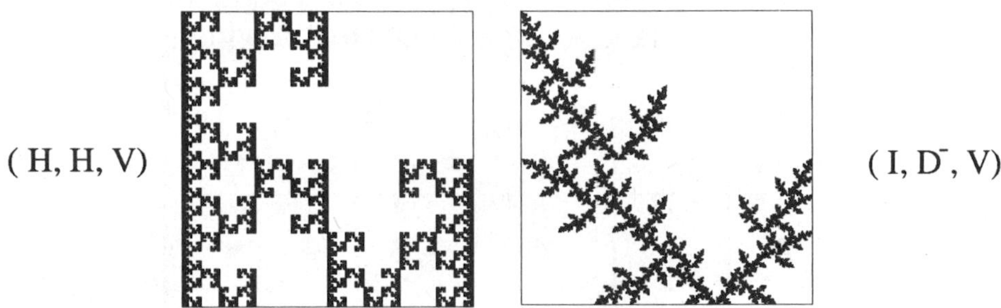

In the generating processes for these fractals, both are reduced and replicated in the same way.The only difference is in how these reduced copies are embedded back into the square building template used throughout the iterative procedures. The transformations used at this point are revealed through the geometric genetic code. This activity focuses on developing a model that contains all the components needed to characterize a particular iterated function system of this type, including how the square template is divided and which cells are used.

1. The base is determined by the square grid structure into which the reduced copies are embedded at each successive stage. Which of these two grids is used for the fractals shown above?

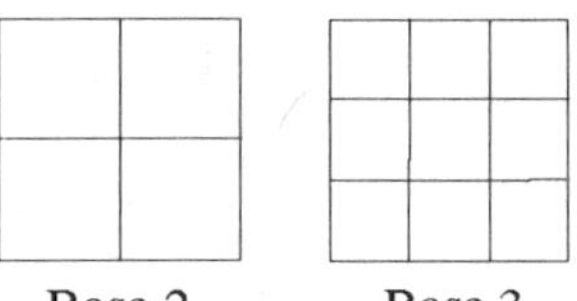

2. How many reduced copies are used at each stage in generating these fractals?
3. What is the reduction scale used at each stage?

Think of the structure as lying in a square. Then imagine the reduction to half size as a dilation of the whole structure by a scale factor of 1/2 toward the center point of the square grid. The three copies are then offset from this central position.

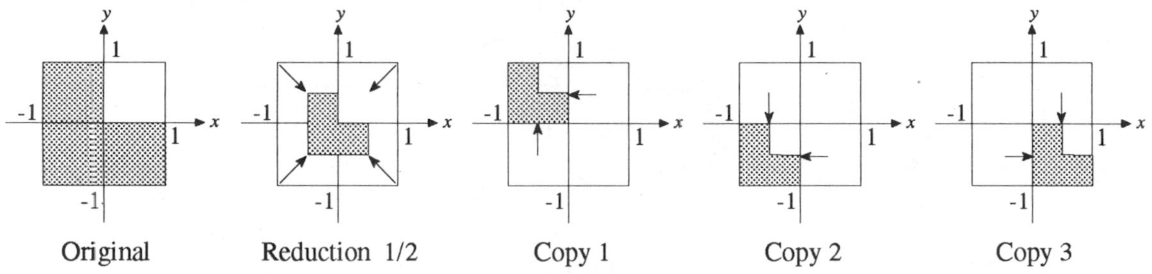

4. Which of the two fractal images shown at the top of the page will emerge from repeated application of this process?

8.6B

One way to summarize this process is through a table that shows the base and the scale, the horizontal and vertical offsets, and the oriewntation for each copy.

5. Enter the appropriate offsets for the three reduced copies needed in this table for generating the Sierpinski triangle. Move copy 1 to the upper left, move copy 2 to the lower left, and move copy 3 in the lower right cell in the underlying grid structure.

Base	Copy	Orientation	Scale	Offset Hori	Offset Vert
2	1	I	1/2		
	2	I	1/2		
	3	I	1/2		

6. The different choices of orientations for the three copies are the eight symmetries of the square. For the Sierpinski triangle, all three entries are I's. What entries are needed in the IFS of the fractal shown on the right on the previous page?

$$I, R_{90}, R_{180}, R_{270}, H, V, D^+, D^-$$

7. Verify that the entries in this IFS table match the fractal image shown.

Base	Copy	Orientation	Scale	Offset Hori	Offset Vert
2	1	V	1/2	-1/2	1/2
	2	R_{270}	1/2	-1/2	-1/2
	3	D^-	1/2	1/2	-1/2

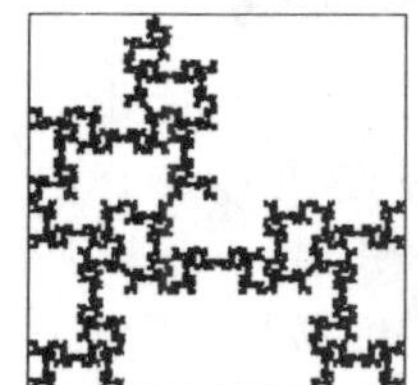

We can see that the entire mechanism of the iterated function system is encoded in the table. Thus, we have a vehicle to analytically present the IFS process that is equivalent to showing the actual geometric figure itself.

Draw stages 1, 2, and 3 of the emerging fractal using the IFS code given.

8.

Base	Copy	Orientation	Scale	Offset Hori	Offset Vert
2	1	I	1/2	-1/2	1/2
	2	R_{90}	1/2	-1/2	-1/2
	3	R_{180}	1/2	1/2	-1/2

Stage 1

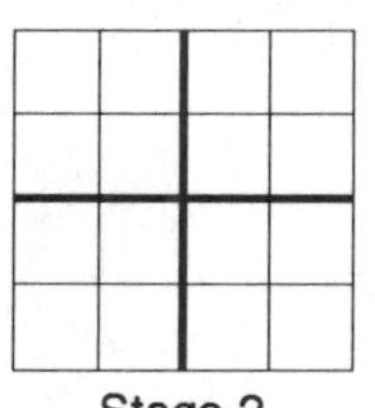

Stage 2

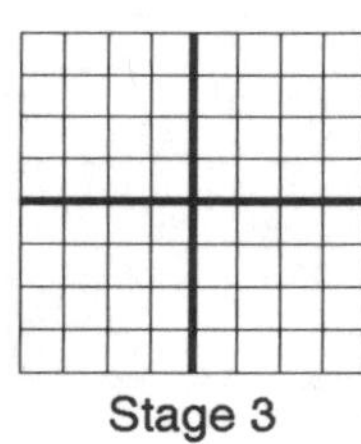

Stage 3

9.

Base	Copy	Orientation	Scale	Offset Hori	Offset Vert
2	1	R_{180}	1/2	-1/2	1/2
	2	R_{90}	1/2	-1/2	-1/2
	3	I	1/2	1/2	-1/2

Stage 1 Stage 2 Stage 3

8.6C

10. Questions 8 and 9 both use the symmetries I, R_{90} and R_{180}. Only their order is different. In how many differnent ways can these three symmetries be assigned to the three copies? Will all the corresponding fractal structures be different?

Match the two IFS codes to the correct fractals shown. Then construct and complete the IFS tables for the remaining two fractals shown.

11. 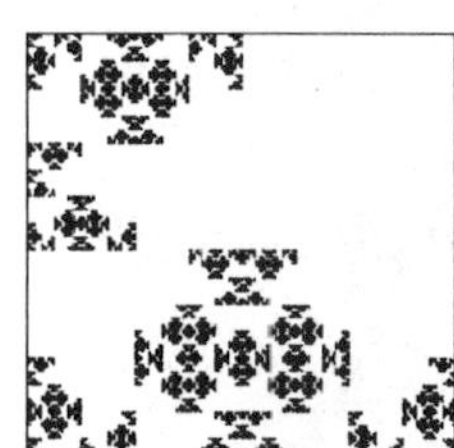

12. 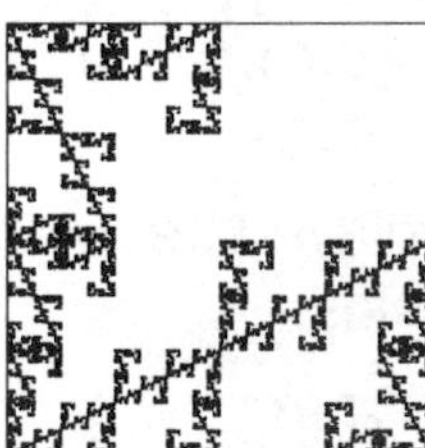

13. 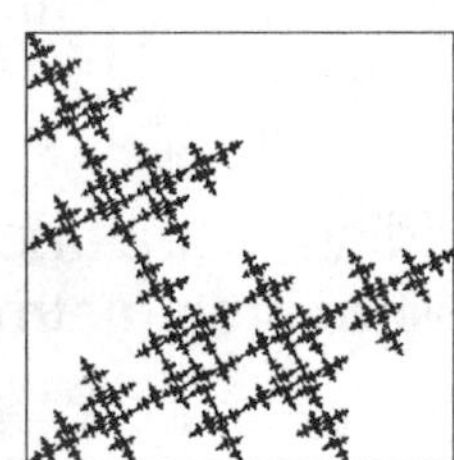

14. 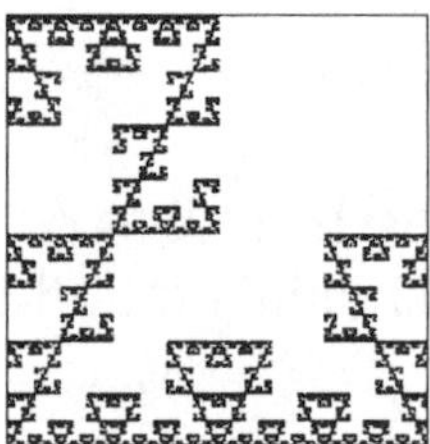

Code for _____:

Base	Copy	Orient-ation	Scale	Offset Hori	Offset Vert
2	1	H	1/2	-1/2	1/2
	2	R_{90}	1/2	-1/2	-1/2
	3	D^+	1/2	1/2	-1/2

Code for _____:

Base	Copy	Orient-ation	Scale	Offset Hori	Offset Vert
2	1	R_{180}	1/2	-1/2	1/2
	2	I	1/2	-1/2	-1/2
	3	V	1/2	1/2	-1/2

Code for _____:

Base	Copy	Orient-ation	Scale	Offset Hori	Offset Vert
2	1		1/2	-1/2	1/2
	2		1/2	-1/2	-1/2
	3		1/2	1/2	-1/2

Code for _____:

Base	Copy	Orient-ation	Scale	Offset Hori	Offset Vert
2	1		1/2	-1/2	1/2
	2		1/2	-1/2	-1/2
	3		1/2	1/2	-1/2

8.6D

Activity 8.3 indicated that each symmetry of the square has an associated matrix representation.

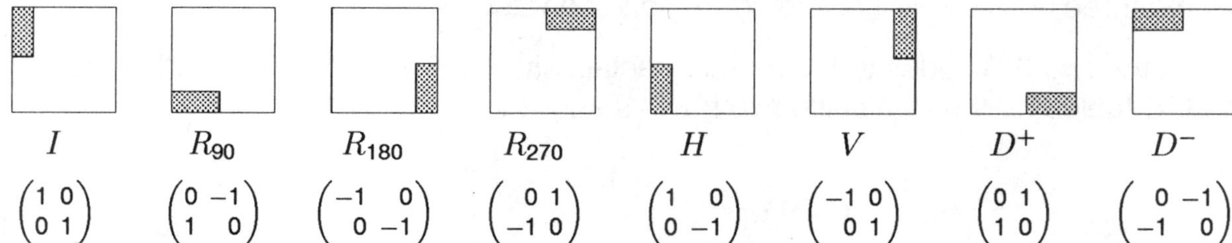

$$I\ \begin{pmatrix}1&0\\0&1\end{pmatrix}\quad R_{90}\ \begin{pmatrix}0&-1\\1&0\end{pmatrix}\quad R_{180}\ \begin{pmatrix}-1&0\\0&-1\end{pmatrix}\quad R_{270}\ \begin{pmatrix}0&1\\-1&0\end{pmatrix}\quad H\ \begin{pmatrix}1&0\\0&-1\end{pmatrix}\quad V\ \begin{pmatrix}-1&0\\0&1\end{pmatrix}\quad D^{+}\ \begin{pmatrix}0&1\\1&0\end{pmatrix}\quad D^{-}\ \begin{pmatrix}0&-1\\-1&0\end{pmatrix}$$

The IFS table corresponding to the fractal displayed below and with problem 7 translates into the following three matrix formulas.

Base	Copy	Orientation	Scale	Offset Hori	Offset Vert
2	1	V	1/2	-1/2	1/2
	2	R_{270}	1/2	-1/2	-1/2
	3	D^-	1/2	1/2	-1/2

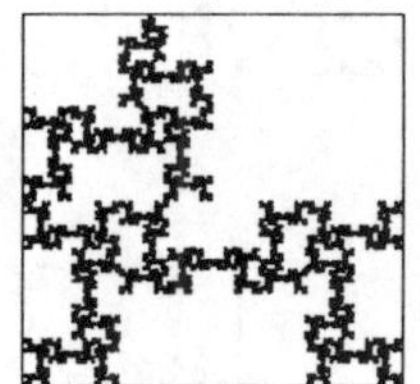

Copy 1: $\begin{pmatrix}x'\\y'\end{pmatrix} = \frac{1}{2}\begin{pmatrix}-1&0\\0&1\end{pmatrix}\begin{pmatrix}x\\y\end{pmatrix} + \begin{pmatrix}-1/2\\1/2\end{pmatrix}$

Copy 2: $\begin{pmatrix}x'\\y'\end{pmatrix} = \frac{1}{2}\begin{pmatrix}0&1\\-1&0\end{pmatrix}\begin{pmatrix}x\\y\end{pmatrix} + \begin{pmatrix}-1/2\\-1/2\end{pmatrix}$

Copy 3: $\begin{pmatrix}x'\\y'\end{pmatrix} = \frac{1}{2}\begin{pmatrix}0&-1\\-1&0\end{pmatrix}\begin{pmatrix}x\\y\end{pmatrix} + \begin{pmatrix}1/2\\-1/2\end{pmatrix}$

15. Complete the three matrix formulas associated with the fractal image shown in problem 11.

Copy 1: $\begin{pmatrix}x'\\y'\end{pmatrix} = \frac{1}{2}\begin{pmatrix}\quad&\quad\\\quad&\quad\end{pmatrix}\begin{pmatrix}x\\y\end{pmatrix} + \begin{pmatrix}-1/2\\1/2\end{pmatrix}$

Copy 2: $\begin{pmatrix}x'\\y'\end{pmatrix} = \frac{1}{2}\begin{pmatrix}0&-1\\1&0\end{pmatrix}\begin{pmatrix}x\\y\end{pmatrix} + \begin{pmatrix}\quad\\\quad\end{pmatrix}$

Copy 3: $\begin{pmatrix}x'\\y'\end{pmatrix} = \frac{1}{2}\begin{pmatrix}\quad&\quad\\\quad&\quad\end{pmatrix}\begin{pmatrix}x\\y\end{pmatrix} + \begin{pmatrix}\quad\\\quad\end{pmatrix}$

16. Write the three matrix formulas associated with the fractal image of problem 14.

8.6E

This page summarizes five distinct approaches to representing an iterated function system.

A. FLOW CHART

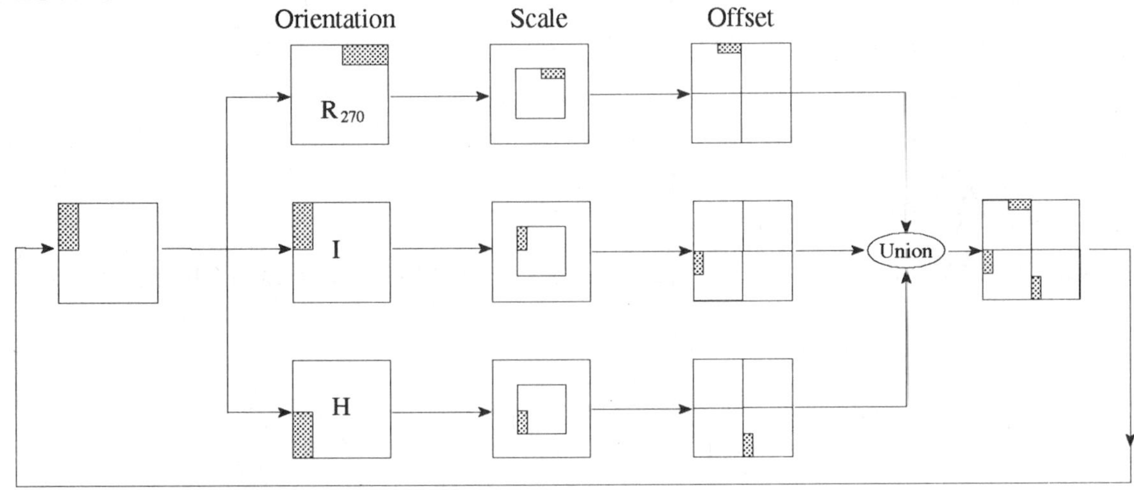

B. GEOMETRIC GENETIC CODE

R_{270}	
I	H

or (R_{270}, I, H)

C. TABLE REPRESENTATION

Base	Copy	Orientation	Scale	Offset Hori	Offset Vert
	1	R_{270}	1/2	-1/2	1/2
2	2	I	1/2	-1/2	-1/2
	3	H	1/2	1/2	-1/2

D. MATRIX REPRESENTATION

Copy 1: $$\begin{pmatrix} x' \\ y' \end{pmatrix} = \frac{1}{2}\begin{pmatrix} 0 & 1 \\ -1 & 0 \end{pmatrix}\begin{pmatrix} x \\ y \end{pmatrix} + \begin{pmatrix} -1/2 \\ 1/2 \end{pmatrix}$$

Copy 2: $$\begin{pmatrix} x' \\ y' \end{pmatrix} = \frac{1}{2}\begin{pmatrix} 1 & 0 \\ 0 & 1 \end{pmatrix}\begin{pmatrix} x \\ y \end{pmatrix} + \begin{pmatrix} -1/2 \\ -1/2 \end{pmatrix}$$

Copy 3: $$\begin{pmatrix} x' \\ y' \end{pmatrix} = \frac{1}{2}\begin{pmatrix} 1 & 0 \\ 0 & -1 \end{pmatrix}\begin{pmatrix} x \\ y \end{pmatrix} + \begin{pmatrix} 1/2 \\ -1/2 \end{pmatrix}$$

E. ATTRACTOR IMAGE

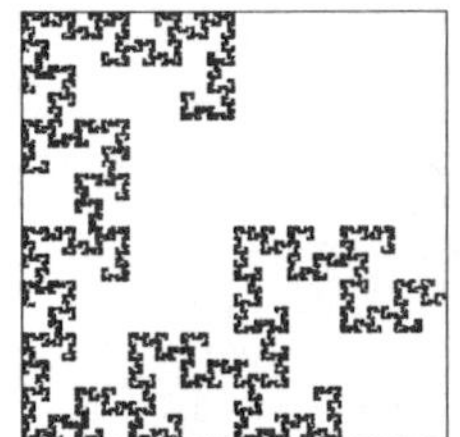

8.6F

A variation of the preceding process could involve a 3 × 3 grid.

17. Complete a table implementation of the attractor shown.

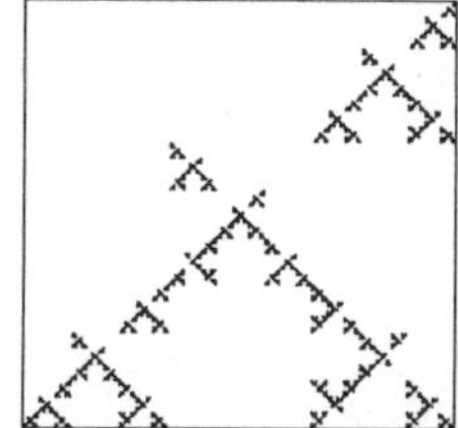

Base	Copy	Orient-ation	Scale	Offset Hori	Offset Vert
	1				
	2				
3	3				
	4				

18. Represent the same iterated function system using the flow chart form and the matrix form.

8.7 CHAOS GAME VARIATIONS 8.7A

We have been studying a special class of fractals based on the eight symmetries of the square. Once the geometric genetic code for one of these fractals is known, enough complete stages can be drawn to develop a fairly good visual image of the final result. But the process of building from one stage to the next can be tedious and time consuming. An entirely different process can be utilized to generate the fractal image on a computer or calculator screen using the method developed in the chaos game.

The essence of this process centers around drawing a single point at each stage rather than the complete figure. The speed of iteration is limited only by the technology being used. Many points can be generated quickly. It is the collection of a large set of such points that displays the desired fractal image. The surprise is that a well ordered fractal emerges from a completely random process of choosing which of the three subsquares will be used to locate the point at each and every stage.

This example shows the two different methods applied to generating the fractal with the geometric genetic code (V, R_{90}, D^{-}). Compare the first four stages in the step-by-step development with the images from 50, 500, 5000, and 50,000 successive points plotted through the appropriately modified chaos game on a computer screen.

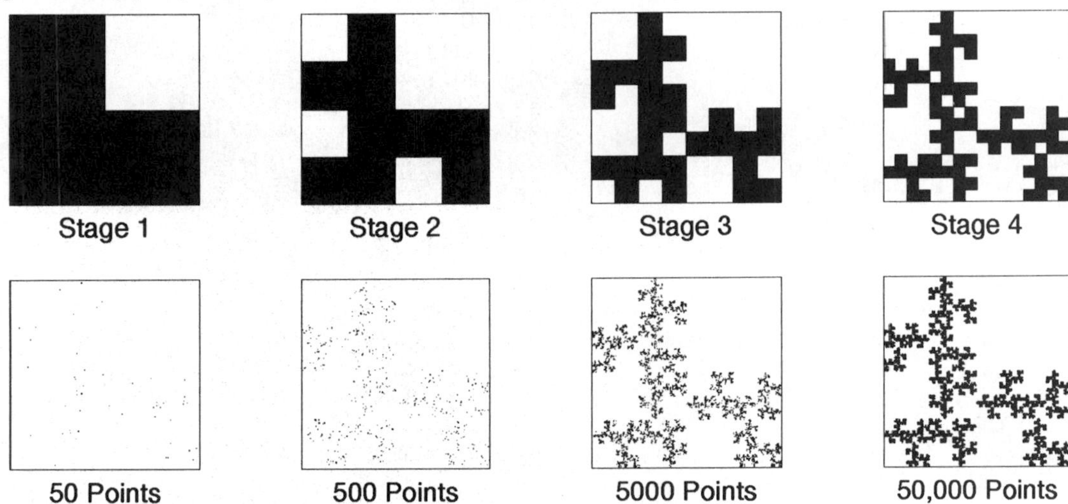

Stage 1 Stage 2 Stage 3 Stage 4

50 Points 500 Points 5000 Points 50,000 Points

1. Does the general shape of the fractal begin to appear at stage 4 in the building sequence? When does it clearly emerge using the chaos game method?

2. Give the three sets of linear equations needed in the program to generate the fractal with the genetic code (V, R_{90}, D^{-}). Give the corresponding matrix equations.

3. Is the image generated from 50,000 points the final fractal or simply some stage in its development? Explain your answer.

4. Does the image generated from 50,000 points represent true self-similarity? Are the three components in cells a, b, and c exact replicas of the whole?

8.7B

5. A graphing calculator screen measures 96×64 pixels. If the initial square measures 64×64 pixels, what is the pixel size of the stage-1 subsquares? What is the pixel size of the subsquares at stage-2, stage-3 and stage-4?

6. At what stage do the squares of the grid cells become smaller than the pixel size? Explain how this affects the detail possible on the screen.

These graphing calculator programs use a modified chaos game to display the fractal for (V, R_{90}, D^-). Simple changes can be made to generate other fractals of this type.

CASIO

```
Range −1.33,1.33,.1, −1,1, .1
"X="? → P
"Y="? → Q
0 → K
Lbl 1
K+1 → K
P → X
Q → Y
Plot X,Y
K > 1000 => End
Ran# → N
N < .3333 => Goto 2
N > .6666 => Goto 4
Goto 3
Lbl 2: 'CELL A'
−X → P
Y → Q
.5P − .5 → P
.5Q + .5 → Q
Goto 1
Lbl 3: 'CELL B'
−Y → P
X → Q
.5P − .5 → P
.5Q − .5 → Q
Goto 1
Lbl 4: 'CELL C'
−Y → P
−X → Q
.5P + .5 → P
.5Q − .5 → Q
Goto 1
```

TEXAS INSTRUMENTS

```
:ClrDraw
:−1.33 → Xmin
:.1.33 → Xmax
:−1 → Ymin
:1 → Ymax
:Disp "X?"
:Input P
:Disp "Y?"
:Input Q
:0 → K
:Lbl 1
:K+1 → K
:P → X
:Q → Y
:Pt-On(X,Y)
:If K > 1000
:End
:rand → N
:If N < .3333
:Goto 2
:If N > .6666
:Goto 4
:Goto 3
:Lbl 2
:−X → P
:Y → Q
:.5P − .5 → P
:.5Q + .5 → Q
:Goto 1
:Lbl 3
:−Y → P
:X → Q
:.5P − .5 → P
:.5Q − .5 → Q
:Goto 1
:Lbl 4
:−Y → P
:−X → Q
:.5P + .5 → P
:.5Q − .5 → Q
:Goto 1
```

8.7C

7. Key in the program. Run the 1000 iterations starting with the point (0,0). Compare the results with those on the previous page for the same fractal (V, R_{90}, D^-).

8. Run the program several times again with different starting points within the range defined in the program. Does essentially the same image appear each time? What happens when you increase the number of iterations in line 10 from 1000 to 2000?

9. Study the program. If the random number generated in line 11 is .4567, what transformation will be applied and in which cell will the point be plotted? Answer the same questions for the random numbers .6789 and .2322.

A genetic code of (V, R_{90}, D^-) indicates that a reflection about the vertical axis occurs in cell A. The symmetry transformation equations for cell A have been used in writing lines 25 and 26 of the program. The complete transformation equations for cell A have been used in writing lines 25 through 28. They involve not only this reflection but the reduction and relocation as well, as shown below. Note the new location of the point. Remember, this point is all that is plotted for this iteration in the chaos game process.

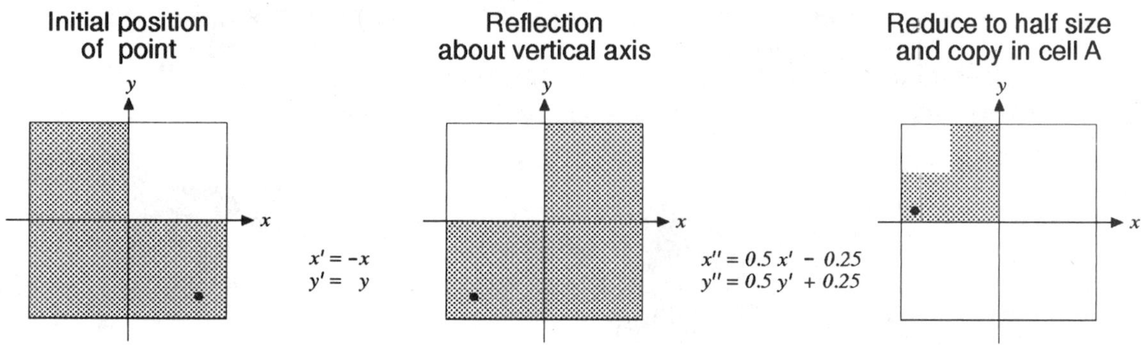

10. Lines 22 and 23 of the program come from the mapping equations shown below. What symmetry transformations of the square do they represent?

Line 32 −y→ P	mapping x' = − y
Line 33 x → Q	mapping y' = x

11. What symmetry transformation equations were used in writing lines 37 and 38?

12. Replace lines 37 and 38 with the two entries in lines 25 and 26 and run the program again. The new fractal generated should have the code (V, D^-, V).

13. Make all changes in the program needed to generate the fractal with the code (V, V, V). Run the program and compare your results with the computer image on the right.

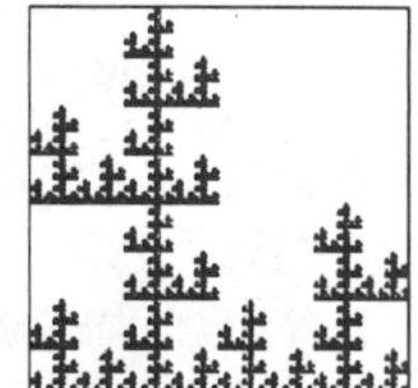

14. How should the initial program be changed in order to generate a fractal where the R_{90} mapping is applied to all three cells? the D^- mapping?

8.7D

15. The Sierpinski triangle is generated when the identity mapping is used in each of the three cells. Modify the program accordingly and run it to verify your changes.

It should now be apparent that lines 25 and 26, lines 31 and 32, and lines 37 and 38 hold the key. By placing the different mappings into these steps, the entire family of fractals based on the eight symmetries of the square can be generated. Here is the complete set of possible entries.

I	R_{90}	R_{180}	R_{270}	H	V	D^+	D^-
x→ P	−y→ P	−x→ P	y→ P	x→ P	−x→ P	y→ P	−y→ P
y→ Q	x→ Q	−y→ Q	−x→ Q	−y→ Q	y→ Q	x→ Q	−x→ Q

Vary the chaos game program to generate these fractals on the graphing calculators.

16. $(R_{270}, R_{180}, R_{90})$ 17. (I, H, D^+) 18. (D^-, D^+, R_{270})

By studying the self-similarity properties of this type of fractal, the corresponding geometric genetic codes can be discovered. Modifications in the calculator program should enable us to reproduce our own images as variations of the chaos game.

Study these fractal images. Identify their geometric genetic codes. Modify the calculator program accordingly, and generate your own copies.

19.

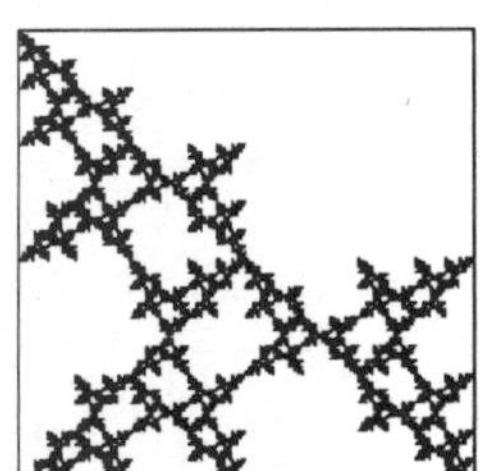

20.

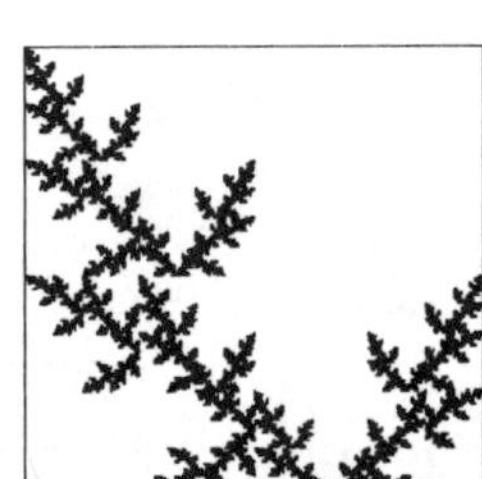

21.

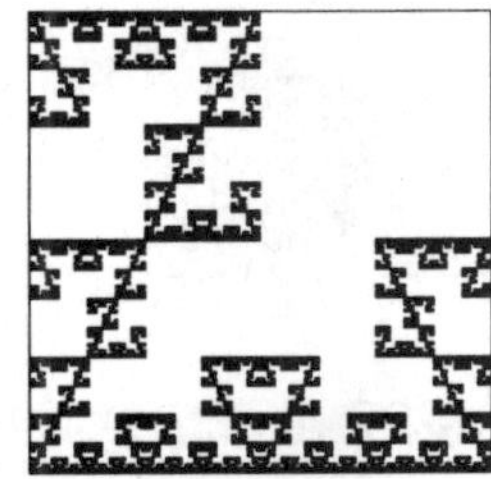

The fractal images shown above were generated on a computer with much higher resolution than that of the 96 × 64 pixels on the graphing calculator screen. You might try to reproduce them yourself on a computer screen.

Much is revealed through the geometric genetic code. It shows

- the self-similarity feature of the final fractal image,
- the building algorithm used to create the fractal through iteration,
- the way sets of different fractals are related through the same building blocks.

The next activity investigates the geometric properties of an entire family of fractals and some of its relatives. The final activity of the unit explores the composition properties for the eight symmetries of the square and uses them to develop a process for finding the building codes for all such relatives from that of a single member.

8.8 FAMILIES OF FRACTALS

8.8A

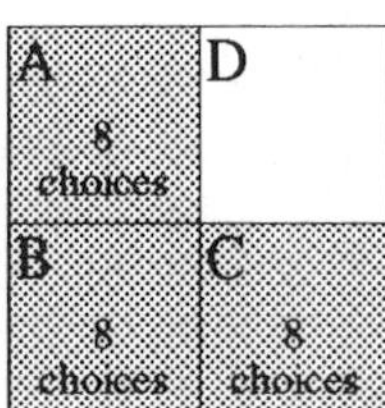

The process described in this unit can be used to produce an entire family of fractals that are close relatives to the Sierpinski triangle. There are eight choices of symmetry transformations in each of the three cells in the array for a total of $8 \times 8 \times 8 = 512$ different geometric genetic codes.

8 transformation choices: I R_{90} R_{180} R_{270} H V D^+ D^-
3 cell choices: Cell A Cell B Cell C

These codes produce an amazing variety of fractals with widely different patterns and forms. Only eight figures in the set are symmetric, each with codes that occur in eight different forms. All the other fractals have unique codes of their own.

1. This fractal has a line of symmetry that allows it to be generated by eight different building codes. Four are listed here. Based on the properties of the different symmetry transformations given, what would you speculate are the other four building codes?

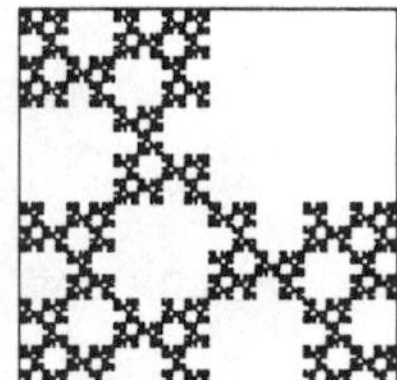

(R_{180}, I, R_{180})
(D^-, I, D^-)
(R_{180}, D^+, R_{180})
(D^-, D^+, D^-)

2. Here is another fractal with a line of symmetry. Motion V or R_{90} can be used for cell A. Motion I or D^+ can be used for cell B. Motion H or R_{270} can be used for cell C. Write the eight different genetic codes that define this fractal.

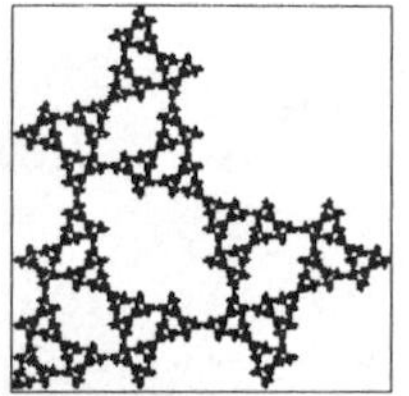

3. Study these three fractals for lines of symmetry. Give either the eight possible choices or the unique genetic code, depending on the presence or absence of a line of symmetry.

a.

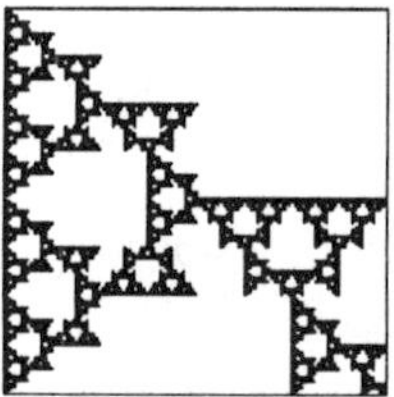

b.

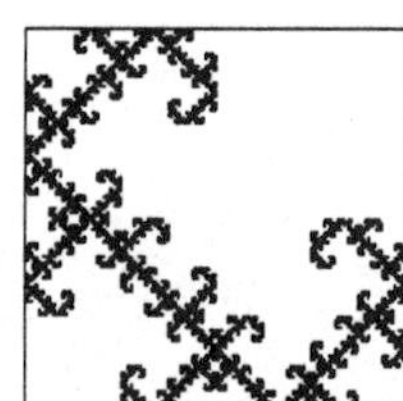

c.

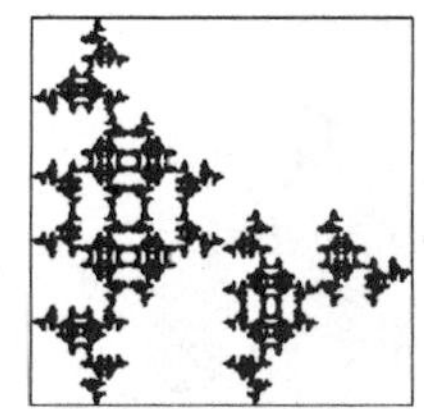

8.8B

The eight fractals with line symmetry can be observed as early as stage 2 in the iterative process of development.

a. b. c. d. e. f. g. h.

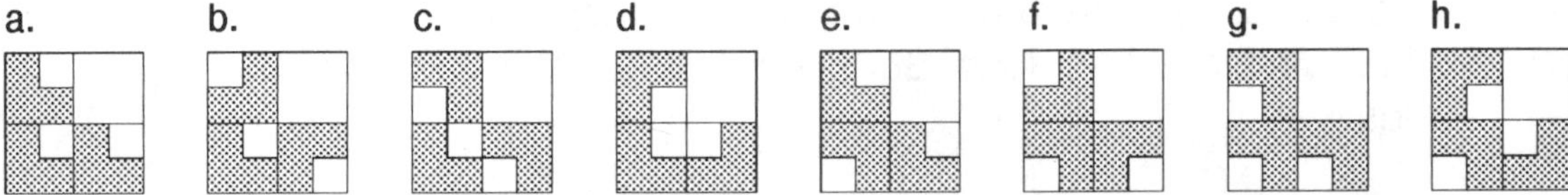

For each fractal, the choices for cell B are I or D^+ in figures a–d and R_{180} or D^- in figures e–h. For each choice, there are always two choices for cells A and C as well. This gives the eight choices shown above.

4. Give at least one building code for each of the eight stage-2 symmetric possibilities shown above.

5. Match each of the symmetric fractals in questions 1, 2, and 3 with the correct stage-2 figure shown above.

6. How many different fractal images occur from the 512 possible genetic codes?

7. Neither of these two fractals has line symmetry, but there is a connection between them. How are they related geometrically?

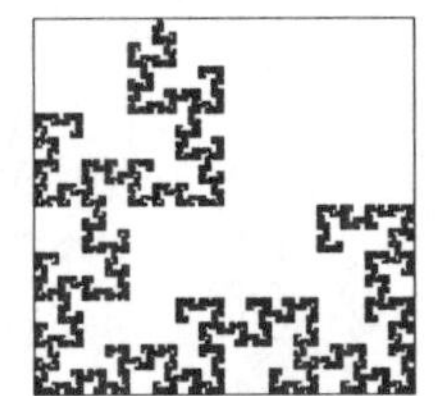

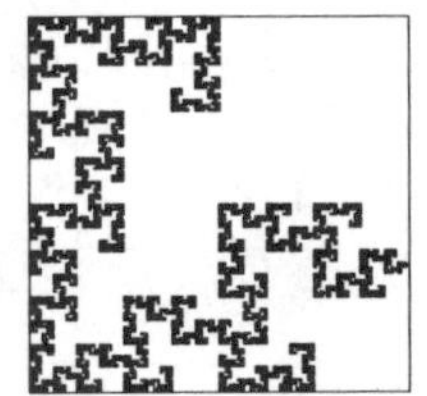

The two fractals above are reflections of one another. The reflection is motion D^+, one of the symmetries of the square. When applied to one figure, the other emerges. All non-symmetric fractals appear in pairs of this type. This raises an interesting question. Is there a whole set of fractal relatives that are related to each other through the eight transformations symmetries? The answer is yes.

8. Assume the first figure below is the identity position. Describe the other relatives as symmetry transformations of the identity position.

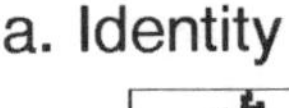

a. Identity b. c. d.

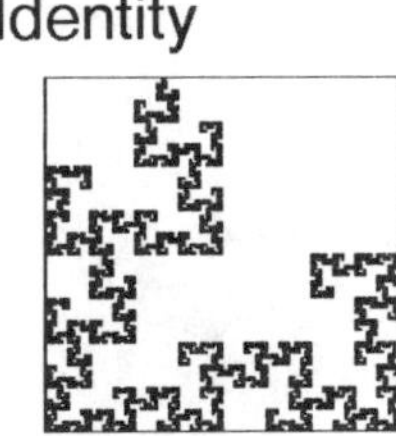

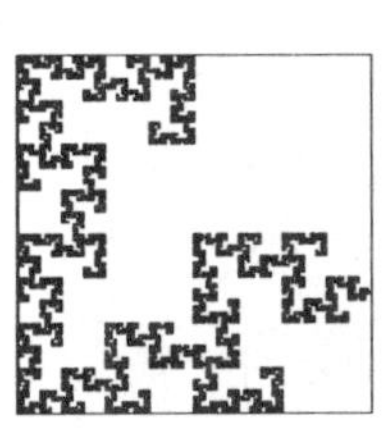

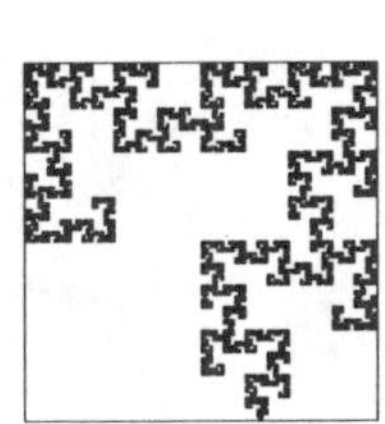

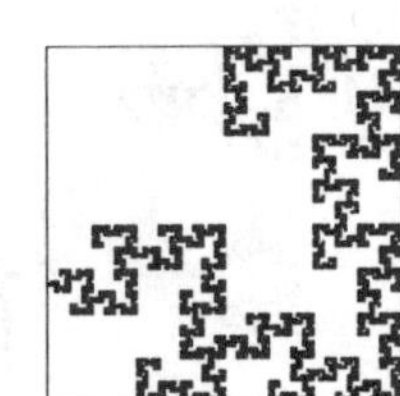

e. f. g. h.

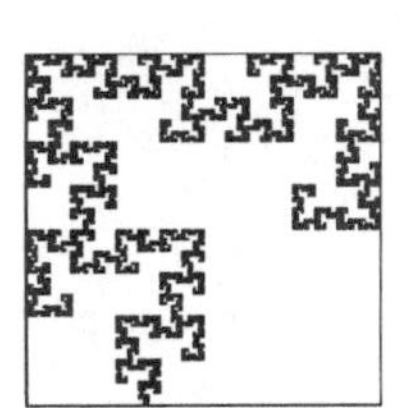

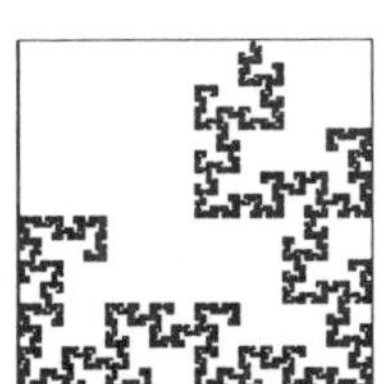

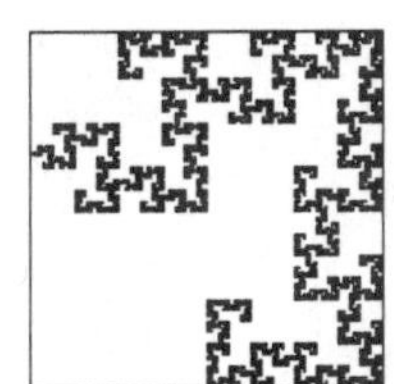

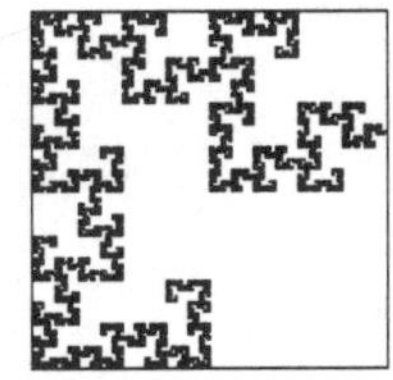

8.8C

The first two fractals above were already counted as close relatives to the Sierpinski triangle. They only occur in cells A, B, and C and thus belong to the family counted on the previous page. The other six are more distant relatives, not mentioned earlier in this unit. They will require identification using all four cells, A, B, C, and D.

9. Imagine the eight symmetry transformations are applied to the square shown. Complete the table of new positions for the four letters P, Q, R, and S.

P	S
Q	R

	Transformations							
	I	R_{90}	R_{180}	R_{270}	H	V	D^+	D^-
Cell A	P	S	R					
Cell B	Q	P	S					
Cell C	R	Q	P					
Cell D	S	R	Q					

If a fractal has line symmetry, then the eight transformations produce only four different images as relatives of this type. They are shown here with the Sierpinski triangle.

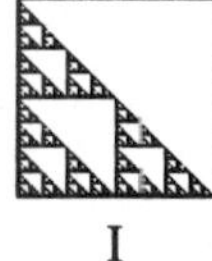
I

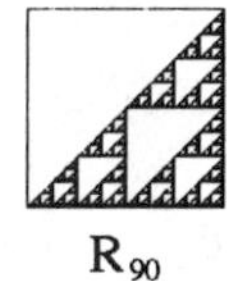
R_{90}

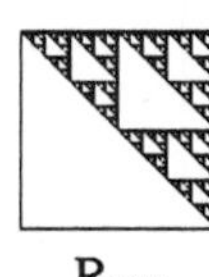
R_{180}

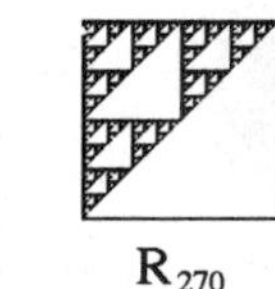
R_{270}

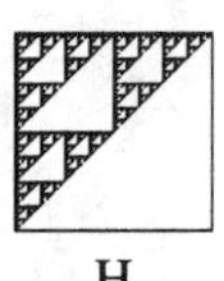
H

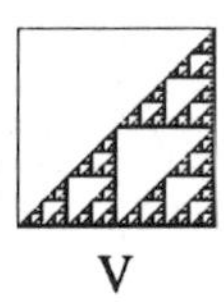
V

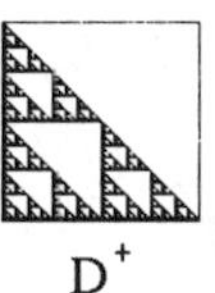
D^+

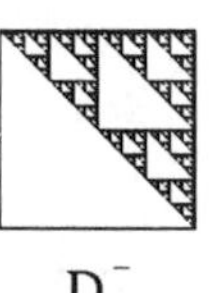
D^-

10. List the transformations that produce matching pairs of figures.

Of particular interest are the transformation codes revealed within the cells of both an original and its relatives formed by rotation or reflection. Examples are shown here for both a symmetric and non-symmetric fractal.

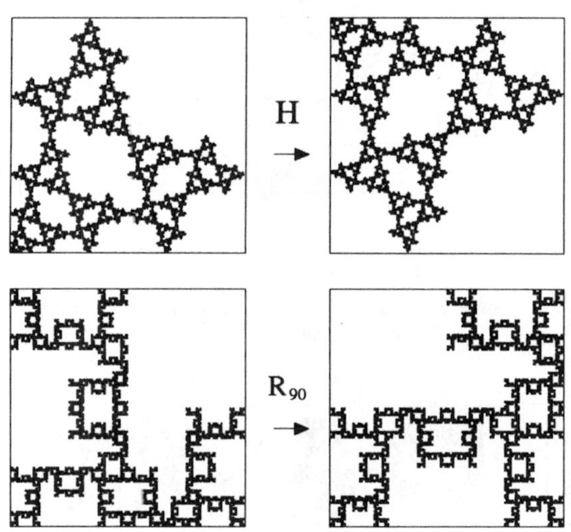

Cell	Codes: Original	Codes: Relative
A	R_{90} or V	I or D^-
B	I or D^+	R_{270} or V
C	R_{270} or H	empty
D	empty	R_{90} or H

Cell	Original	Relative
A	D^-	empty
B	R_{90}	D^+
C	V	R_{90}
D	empty	H

8.8D

Each image below on the left is associated with a relative. The relative is obtained by applying one of the eight symmetries of the square to the original. Complete the corresponding codes in the tables on the right. You may want to cut out and use the manipulative aids at the bottom of the page. Fold each one on the dotted line so that the reduced original image is on one side and its reflection on the other side.

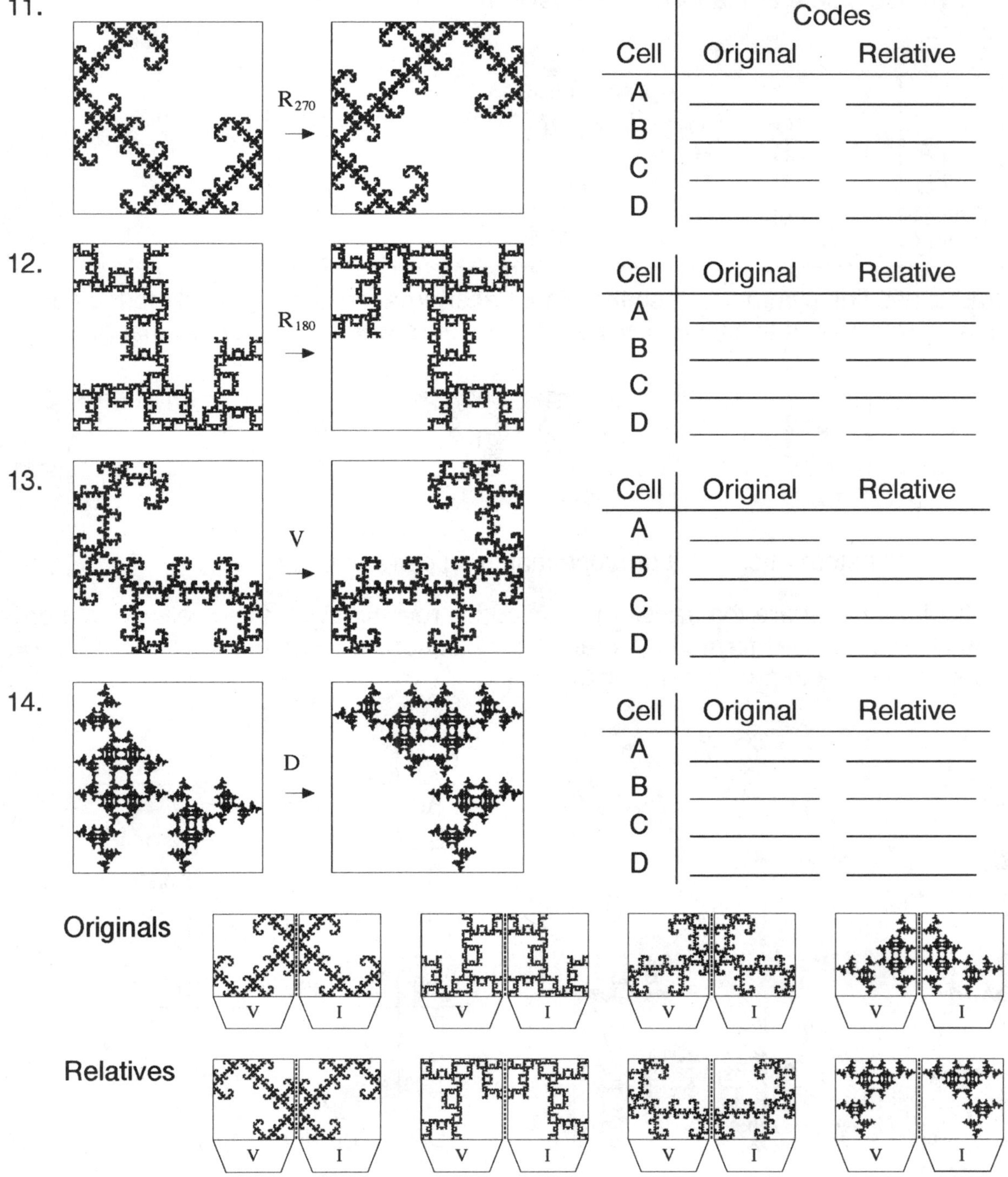

11.

Cell	Codes	
	Original	Relative
A		
B		
C		
D		

12.

Cell	Original	Relative
A		
B		
C		
D		

13.

Cell	Original	Relative
A		
B		
C		
D		

14.

Cell	Original	Relative
A		
B		
C		
D		

8.9 ORIGINALS AND RELATIVES **8.9A**

What tools are needed to identify algebraically the relatives of a fractal? They come from the properties of composition of the symmetries of a square.

$T_2 \circ T_1$		T_2 = second operation							
		I	R_{90}	R_{180}	R_{270}	H	V	D^+	D^-
	I	I	R_{90}	R_{180}	R_{270}	H	V	D^+	D^-
	R_{90}	R_{90}	R_{180}	R_{270}	I	D^-	D^+	H	V
T_1	R_{180}	R_{180}	R_{270}	I	R_{90}	V	H	D^-	D^+
first	R_{270}	R_{270}	I	R_{90}	R_{180}	D^+	D^-	V	H
oper-	H	H	D^+	V	D^-	I	R_{180}	R_{90}	R_{270}
ation	V	V	D^-	H	D^+	R_{180}	I	R_{270}	R_{90}
	D^+	D^+	V	D^-	H	R_{270}	R_{90}	I	R_{180}
	D^-	D^-	H	D^+	V	R_{90}	R_{270}	R_{180}	I

This table exhibits the composition of the eight rotations and reflections of a square. For example, the rotation R_{270} followed by the reflection H gives the same result as the single motion D^+: $H \circ R_{270} = H[R_{270}] = D^+$.

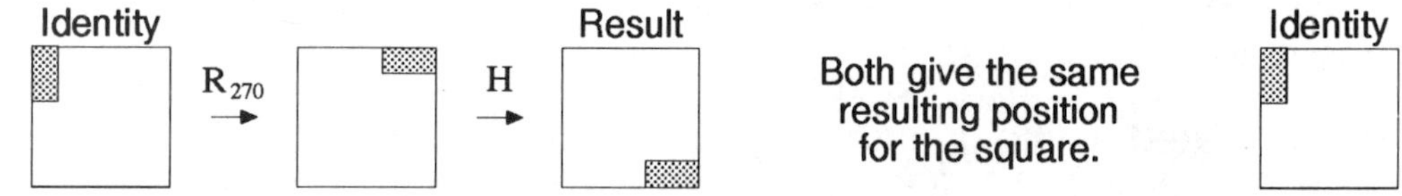

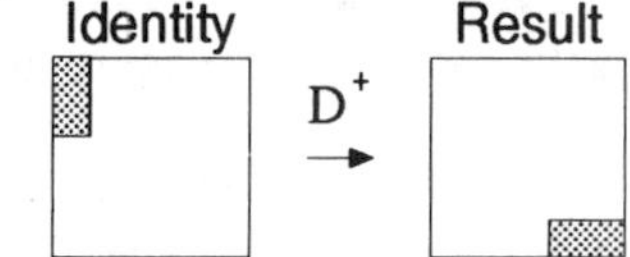

1. Does every composition of two of the eight motions of a square lead to one of the symmetries of the square?. Use the table to find the single motion equivalent to each composition. Draw the two correponding diagrams for

   a. $V \circ R_{90}$ b. $D^- \circ D^+$ c. $R_{180} \circ H$

When the motion H follows that of R_{270}, the result is the same as the single motion D^-. What motion would have to follow H to give the same result? Using the table, we find the motion to be

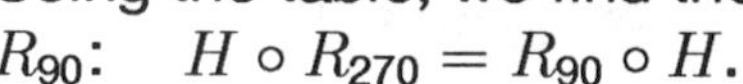

R_{90}: $H \circ R_{270} = R_{90} \circ H$.

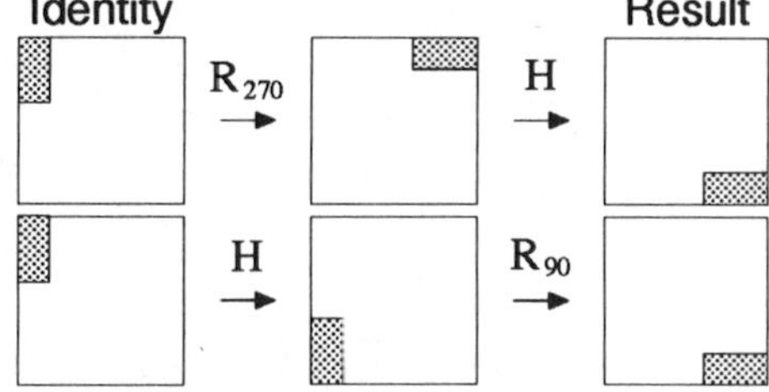

An operation is said to be COMMUTATIVE when reordering the two terms about the operation does not change the results. From the example above, it is clear that H is not commutative with R_{270} or R_{90}. However, H does commute with some motions.

8.9B

2. Find those motions in the composition table that commute with these motions.
   a. H b. R_{90} c. V d. D^-

3. Study this composition table for motions restricted to the three rotations and the identity mapping. Is the composition commutative for this entire subset of mappings?

	I	R_{90}	R_{180}	R_{270}
I	I	R_{90}	R_{180}	R_{270}
R_{90}	R_{90}	R_{180}	R_{270}	I
R_{180}	R_{180}	R_{270}	I	R_{90}
R_{270}	R_{270}	I	R_{90}	R_{180}

Two elements in the composition table are INVERSES of each other if their composition gives the identity motion. Every one of the eight symmetry motions of the square has a unique inverse motion.

4. Find the inverse for each motion using the composition table for all eight mappings shown on the previous page. How many motions act as their own inverses?
   a. I b. R_{90} c. R_{180} d. R_{270}
   e. H f. V g. D^+ h. D^-

How do the properties of composition connect to the identification of fractal relatives? The answer to this question lies in what we know about the geometric nature of fractal relatives and in the notion of conjugation.

Start with the fractal created by the geometric genetic code (D^-, R_{90}, V). Consider its relative formed by the rotation R_{90}, a 90° counterclockwise rotation of the square.

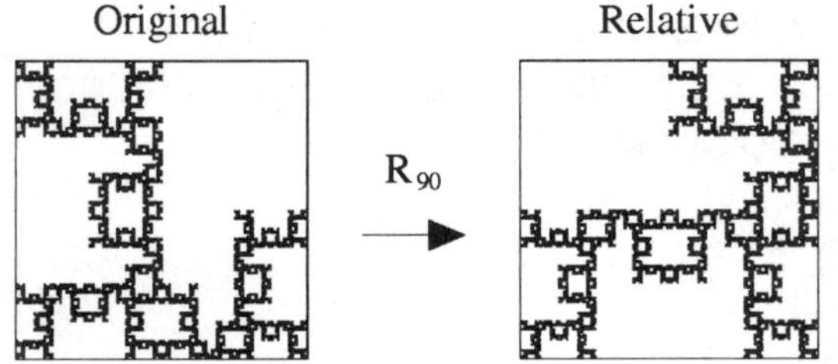

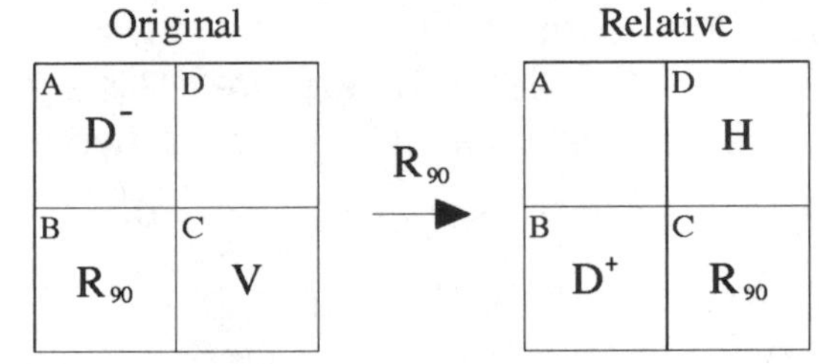

Through the mapping of R_{90}, these changes occur:

Code D^- in cell A becomes code D^+ in cell B.
Code R_{90} in cell B becomes code R_{90} in cell C.
Code V in cell C becomes code H in cell D.

We can see these connections geometrically by looking at the actual results of this rotation on the fractal . But how can we find the changes in the codes of the individual cells algebraically when the motion R_{90} is applied to the entire fractal?

8.9C

The key idea for discovering the algebraic solution to the problem of determining the genetic code of the relative for a given fractal of our family can be motivated by geometric reasoning as illustrated in this flow chart and explained below.

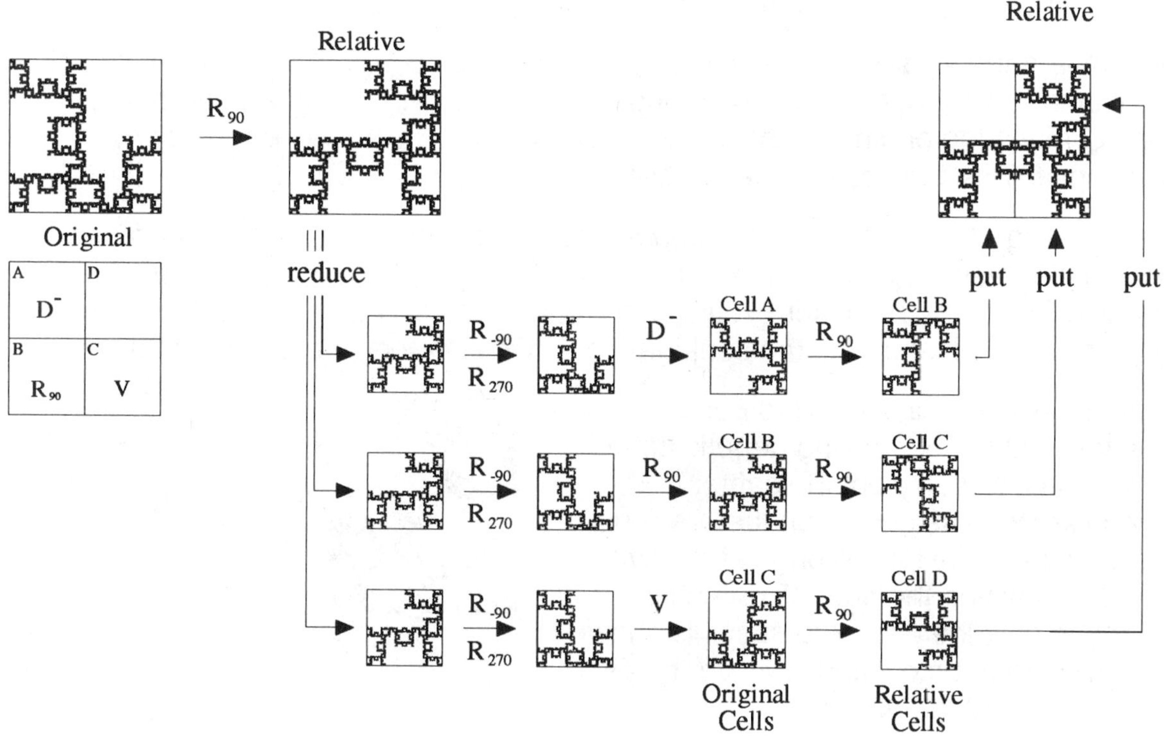

This chart shows how the relative is subdivided into 4 quadrants (one of which is empty) and how each of the pieces can be seen as a transformed copy of a reduced version of the relative, thus, revealing the proper geometric genetic code of the fractal. Remember that the sequence of operations for producing one of the cells of a fractal is first a reduction by a factor of 1/2, then one of the eight standard transformations of the square, and finally a translation to move the piece into its proper position. Thus, the first step is reducing the whole image of the relative, once for each of the three occupied cells. Now, in order to make use of the knowledge of the genetic code of the *original,* we first have to get back to its orientation. In other words, for each of the three small copies we apply the *inverse* operation $R_{-90} = R_{270}$ of the transformation R_{90} that yielded the relative. Then we know that, for example, application of D^- to one of these pieces results in cell A of the original fractal. Therefore, let's apply that transformation, and from the definition of the relative we can now deduce that a rotation of R_{90} and putting the resulting piece into cell B must produce a match with cell B of the relative. The net result is that we get the genetic code for cell B as the concatenation of first R_{270}, then D^-, and finally R_{90}, which, taken together, is the same as D^+. The analogeous train of thought applies to the other two occupied cells as shown in the chart.

8.9D

5. Based on the geometric explanation on the previous page, in order to discover the algebraic connections among the cell mappings for the R_{90} relative, we must solve the following equations.

$$R_{90} \circ D^- \circ R_{270} = \_\_ \qquad R_{90} \circ R_{90} \circ R_{270} = \_\_ \qquad R_{90} \circ V \circ R_{270} = \_\_$$

Here R_{270} is the single *inverse* mapping, which takes the relative and maps it back into the original. D^-, R_{90}, V are the mappings for the geometric genetic code of the original, and R_{90} maps the original to the relative. On the right-hand side we obtain the mappings for the the code of the relative.

The mapping D^+ is called the *conjugate* of D^- by the motion R_{90}. That is to say, in forming the relative fractal by the rotation R_{90}, the cell motion D^+ becomes D^- in the new cell. The same behavior occurs with R_{90} and R_{90} and with V and H, so R_{90} is the conjugate of itself and H is the conjugate of V with respect to rotation by 90° .

6. In order to practice the geometric interpretation of deriving genetic codes for relatives carry out the same construction as shown on the previous page, this time for the original/relative pair shown on the right. To simplify shade only the occupied subcells shown in the graph below and indicate the used mappings at the arrows.

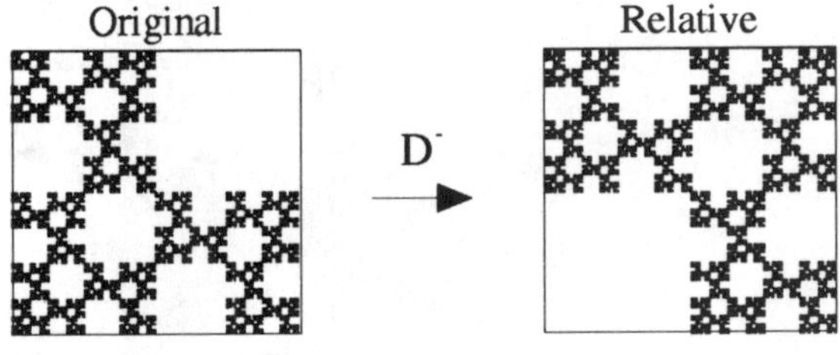

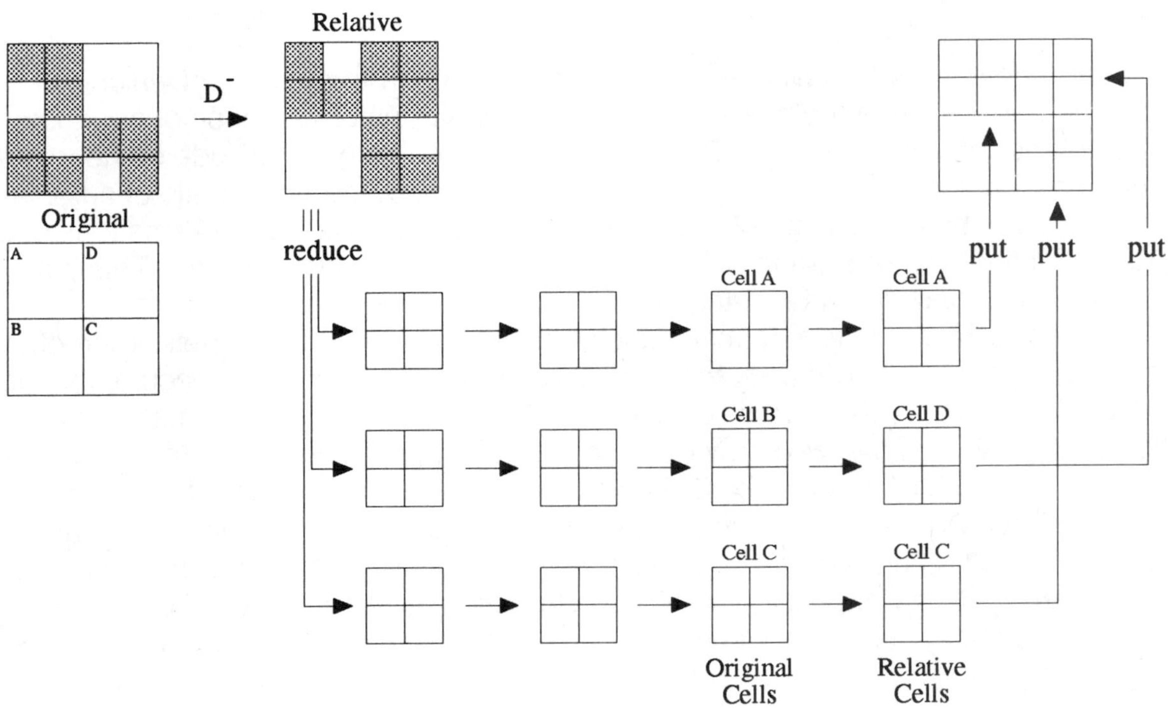

8.9E

In general, to conjugate one of the motions M by another motion m, evaluate the composition $m \circ M \circ m^{-1}$ where m^{-1} is the unique inverse motion associated with m.

7. Use the formula $R_{90} \circ M \circ R_{270}$ to compute the conjugate for each of the eight transformation M by R_{90}. Check to see that your results support those found in question 6 for the relative R_{90}.

The equation $R_{90} \circ D^- \circ R_{270} = D^+$ and $R_{90} \circ D^+ \circ R_{270} = D^-$ show that, under the motion of R_{90}, the motions D^+ and D^- are conjugates of each other. In fact, if some transformation A is the conjugate of a transformation B, then also B is the conjugate of A.

8. Give the motion and the conjugate pairs identified by these equations.
   a. $R_{180} \circ V \circ R_{180} = V$ b. $H \circ R_{90} \circ H = R_{270}$
   c. $D^+ \circ R_{270} \circ D^+ = R_{90}$ d. $R_{270} \circ H \circ R_{90} = V$

9. A fractal has the geometric genetic code (R_{90}, I, D^+). Find the corresponding cell mappings for the D^- relative formed by reflecting the original fractal about the upper-left to lower-right diagonal. Then enter into the correct cells the three conjugates found for this reflected relative of the original fractal (R_{90}, I, D^+).

$D^- \circ R_{90} \circ D^+ = D^- \circ \_\_ = \_\_$
$D^- \circ I \circ D^+ = D^- \circ \_\_ = \_\_$
$D^- \circ D^+ \circ D^+ = D^- \circ \_\_ = \_\_$

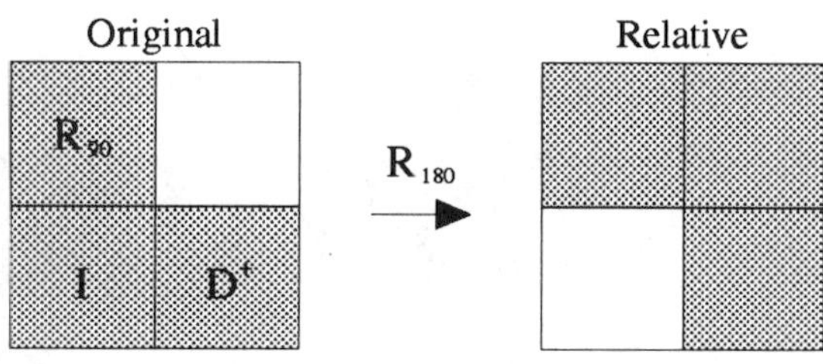

10. Imagine the same original fractal (R_{90}, I, D^+) from question 9, this time being rotated 180° counter-clockwise. Write and solve the corresponding conjugate equations and enter their solutions into the appropriate cells of the R_{180} relative.

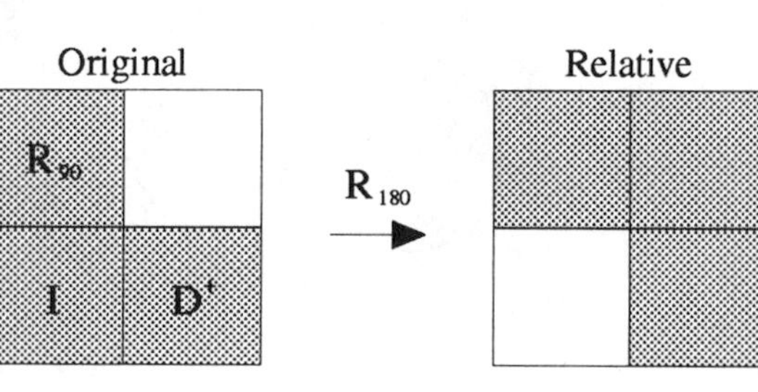

11. Use these equations and the table to compute algebraically the conjugates by H for the three cell mappings given.

Cell	Mapping	Conjugate
A	R_{90}	$H \circ R_{90} \circ H = \_\_$
B	R_{180}	$H \circ R_{180} \circ H = \_\_$
C	V	$H \circ V \circ H = \_\_$

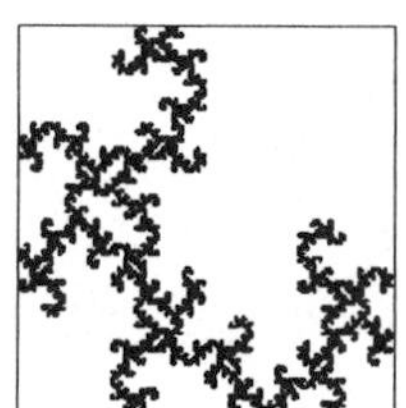

8.9F

The eight symmetries of the fractal shown on the rght and in question 11 are given at the bottom of the page. Which one should be used in cell A of the H relative or will that cell be empty? What about the images in cells B, C, and D?

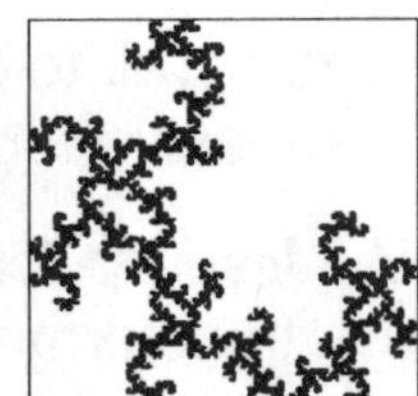

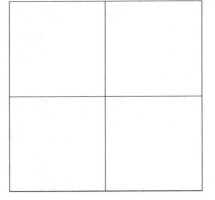

12. Cut out the appropriate images from the bottom of the page and form geometrically the reflection relative H defined algebraically in question 12.

13. The fractal shown in question 12 is one of a set of eight relatives. It is the identity image. The fractal formed in question 13 is one of the reflection relatives. Find algebraically the remaining three reflection and the three rotation relatives. Check the results geometrically by building the corresponding fractals using the appropriate images from those shown below.

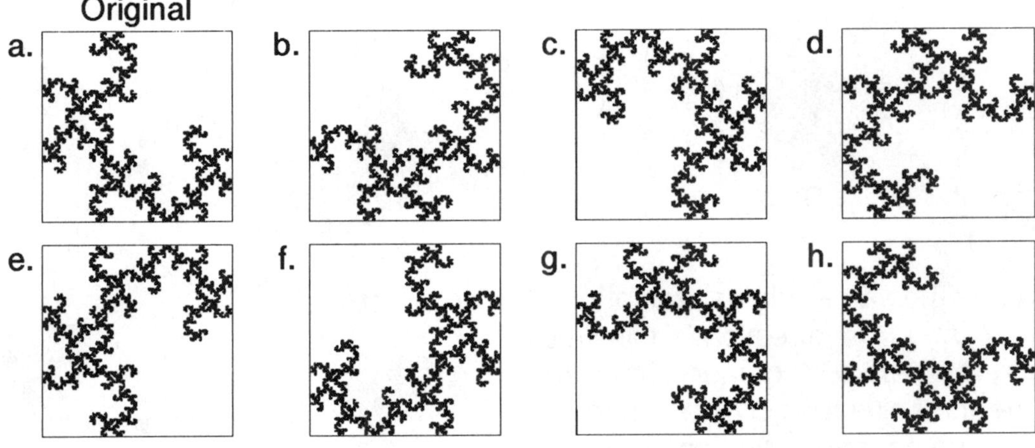

ANSWERS

UNIT 7 IFS in Two Dimensions

ACTIVITY 7.1

3a. flip vertically followed by a 240 degree counterclockwise rotation
3b. no flip , no rotation
3c. flip vertically followed by a 120 degree counterclockwise rotation
4.

5. see the algorithm given in question 4.
6.

Stage	1	2	3	4	n
Segment Length	1	1/2	1/4	1/8	$1/2^{(n-1)}$
Curve Length	3	9/2	27/4	81/8	$3^n/2^{(n-1)}$

6a. $Length(n+1) = Length(n) * 3/2$
6b. $3^n/2^{(n-1)}$
7. 0
8. ∞

ACTIVITY 7.2

1.

Corners: 1 (top left), 3 (top right), 0 (bottom left), 2 (bottom right)

1	3
0	2

11	13	31	33
10	12	30	32
01	03	21	23
00	02	20	22

111	113	131	133	311	313	331	333
110	112	130	132	310	312	330	332
101	103	121	123	301	303	321	323
100	102	120	122	300	302	320	322
011	013	031	033	211	213	231	233
010	012	030	032	210	212	230	232
001	003	021	023	201	203	221	223
000	002	020	022	200	202	220	222

3.

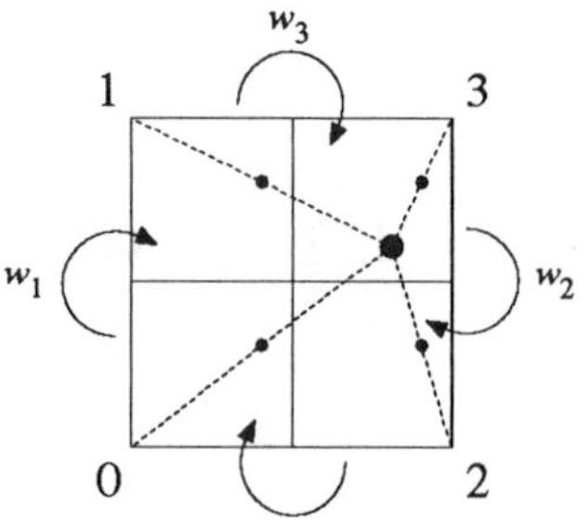

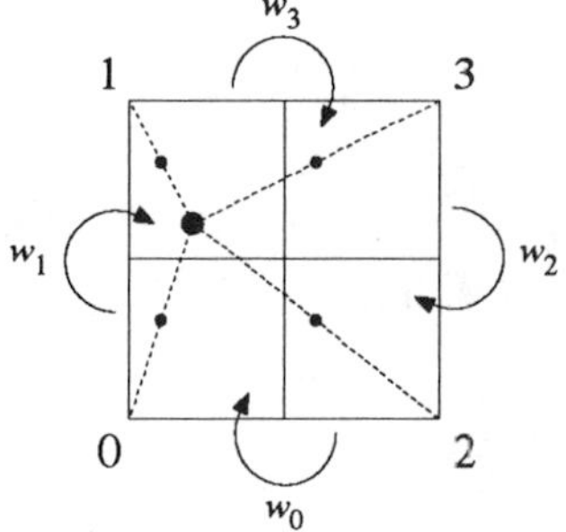

2a.

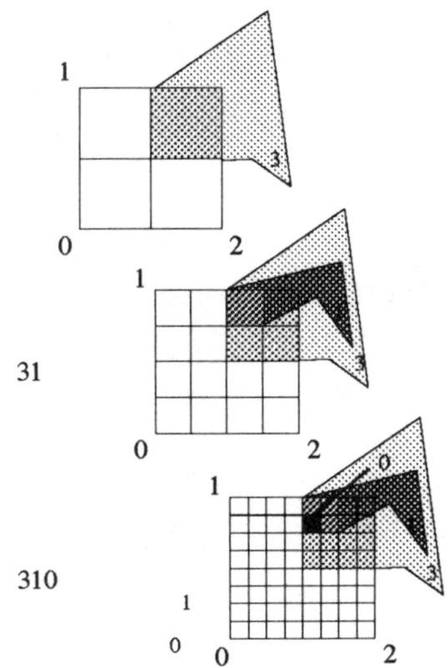

2b.

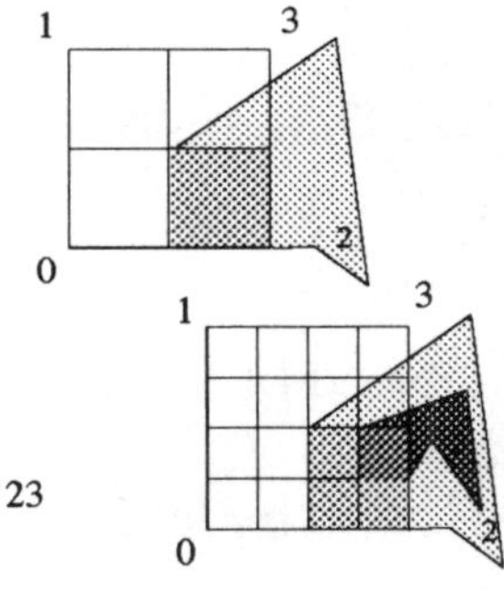

231

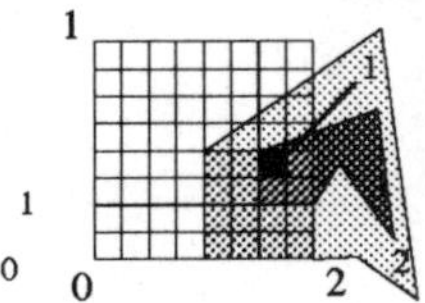

4. ratio 1:1

5a.

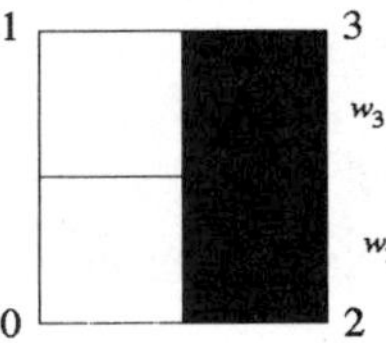

5b.

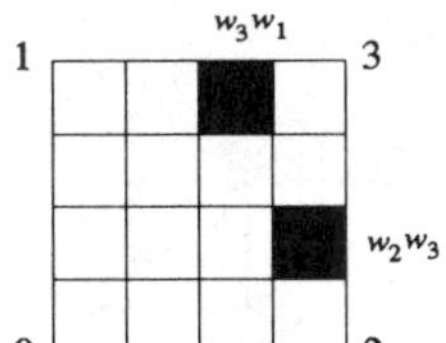

5c.

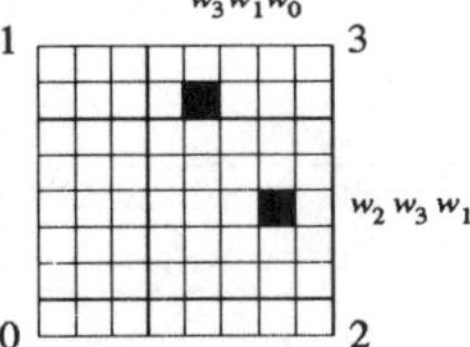

6. It's the same way.

7a.

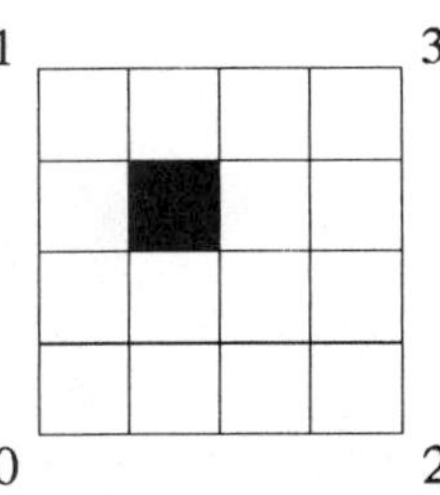

7b.

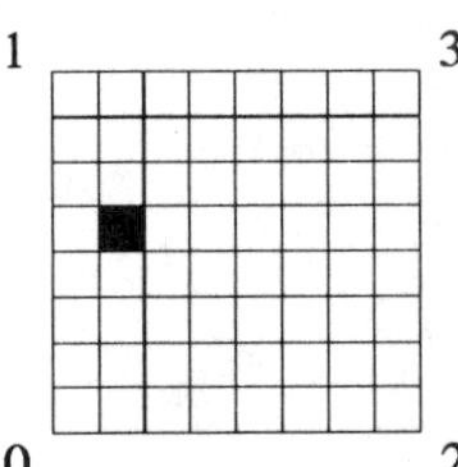

8. n digits are significant.
9. On the next stage four digits are significant.
11.

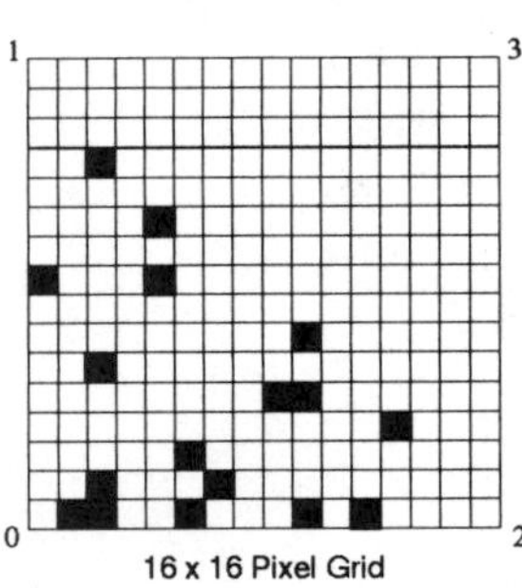

14 a. 3 Pixels b. 9 Pixels c. 27 Pixels d. 3^n Pixels

15. Because there are no addresses which contain a 3.

ACTIVITY 7.3

1.

1	3
0	2

Corner labels: 1 (top left), 3 (top right), 0 (bottom left), 2 (bottom right)

11	13	31	33
10	12	30	32
01	03	21	23
00	02	20	22

Corner labels: 1 (top left), 3 (top right), 0 (bottom left), 2 (bottom right)

2. 221, 3332, 102

3.

3	(0,3)	(1,3)	(2,3)	(3,3)
2	(0,2)	(1,2)	(2,2)	(3,2)
1	(0,1)	(1,1)	(2,1)	(3,1)
0	(0,0)	(1,0)	(2,0)	(3,0)
	0	1	2	3

4.

11	(00,11)	(01,11)	(10,11)	(11,11)
10	(00,10)	(01,10)	(10,10)	(11,10)
01	(00,01)	(01,01)	(10,01)	(11,01)
00	(00,00)	(01,00)	(10,00)	(11,00)
	00	01	10	11

5. Nested 102, Decimal Coord. (1,4), Binary Coord. (001,100)
6. Nested 213, Decimal Coord. (5,3), Binary Coord. (101,011)
7. Nested 012, Decimal Coord. (1,2), Binary Coord. (001,010)
8.

	a.	b.	c.
Decimal Coordinates	(1,4)	$w_1 w_0 w_2$	(001,100)
Binary Coordinates	(6,5)	$w_3 w_2 w_1$	(110,101)
Mapping Composition	(1,2)	$w_0 w_1 w_2$	(001,010)

9.

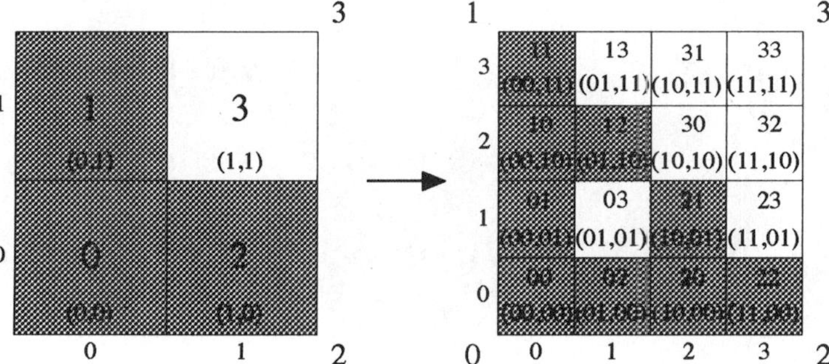

10. Digit 3 fails to appear in the nested addresses associated with shaded.

11.

SHADED CELLS									
Dec. Coords.	(0,0)	(0,1)	(0,2)	(0,3)	(1,0)	(1,2)	(2,0)	(2,1)	(3,0)
Binary Coords.	00 00	00 01	00 10	00 11	01 00	01 10	10 00	10 01	11 00

NON-SHADED CELLS							
Dec. Coords.	(1,1)	(1,3)	(2,2)	(2,3)	(3,1)	(3,2)	(3,3)
Binary	01	01	10	10	11	11	11
Coords.	01	11	10	11	01	10	11

12. The binary coordinates of a shaded cell never have 1's simultaneously appearing in the same column.

ACTIVITY 7.4

1.

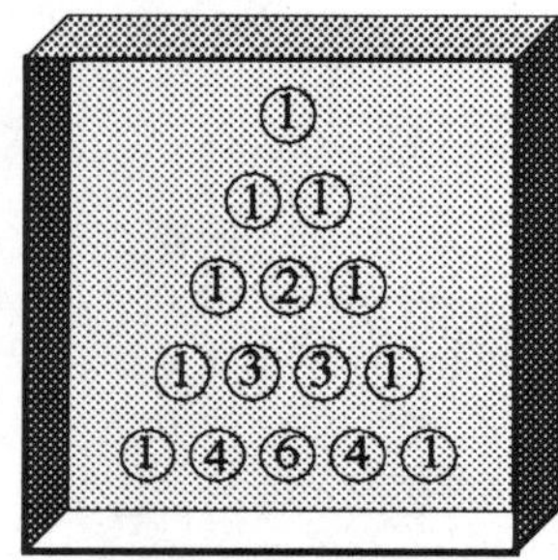

2. Each number of path to a particlar peg can be computed by adding the number in the two pegs above it.

3 a. $\{w\}, \{x\}, \{y\}, \{z\}$ b. $\{w,x\}, \{w,y\}, \{w,z\}, \{x,y\}, \{x,z\}, \{y,z\}$
c. $\{w,x,y\}, \{w,x,z\}, \{w,y,z\}, \{x,y,z\}$ d. $\{w,x,y,z\}$

4 a. 4 b. 6 c. 4 d. 1

5.

1
1 1
1 2 1
1 3 3 1
1 4 6 4 1

6a. The set is {a,b,c,d,e,f,g,h}
{a,b}, {a,c}, {a,d}, {a,e}, {a,f}, {a,g}, {a,h},
{b,c}, {b,d}, {b,e}, {b,f}, {b,g}, {b,h},
{c,d}, {c,e}, {c,f}, {c,g}, {c,h},
{d,e}, {d,f}, {d,g}, {d,h},
{e,f}, {e,g}, {e,h},
{f,g}, {f,h},
{g,h}

6b.

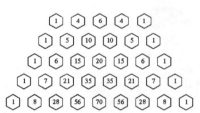

1 4 6 4 1
1 5 10 10 5 1
1 6 15 20 15 6 1
1 7 21 35 35 21 7 1
1 8 28 56 70 56 28 8 1

6c. $\binom{8}{2} = \frac{8!}{2!\cdot 6!} = 28$

ACTIVITY 7.5

1. 4 routes 3 routes 1 route

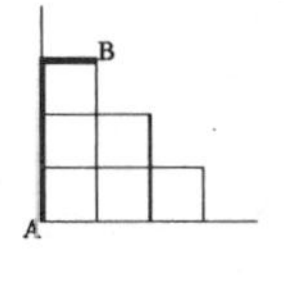

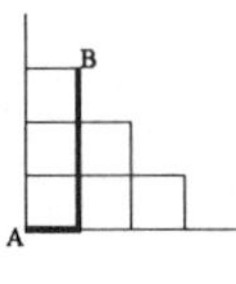

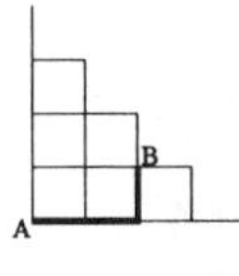

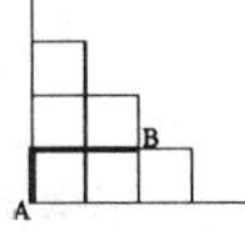

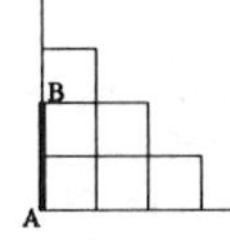

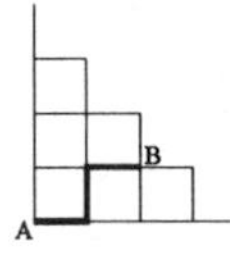

2.

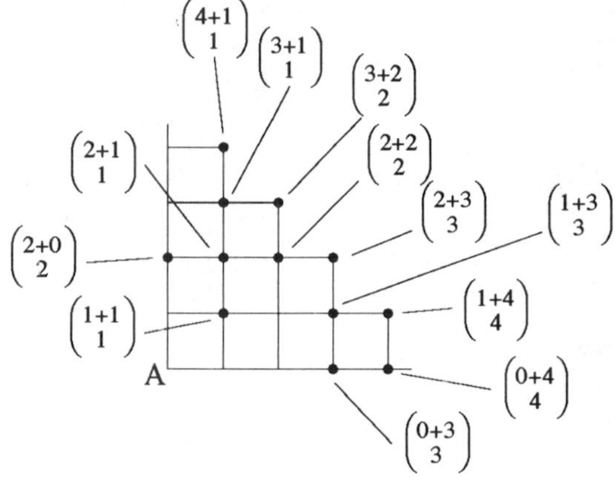

3.

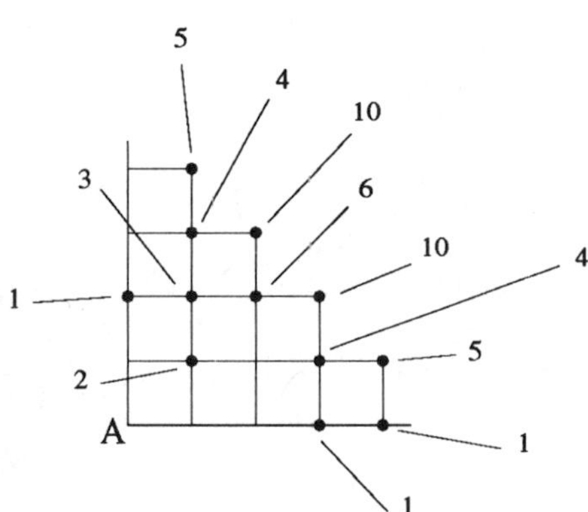

4. The position is horizontal left and vertical under to the location of B.

5a. A $\binom{n+k-1}{k}$, B $\binom{n-1+k}{k}$, X $\binom{n+k}{k}$

5b. $\binom{n+k}{k} = \binom{n-1+k}{k} + \binom{n+k-1}{k-1}$

6. $\frac{(n+k)!}{n!k!} = \frac{(n-1+k)!}{k!(n-1)!} + \frac{(n+k-1)!}{(k-1)!n!} = \frac{(n-1+k)!k+n(n+k-1)!}{k!n!}$

$(n+k)! = (n+k)((n+k-1)!$

7. First multiply the polynom from power $n-1$ with x. Then multiply it with y, then add the two resulting polynoms and get the polynom for power n.

8. $a_k = \binom{m}{k} = \binom{n+k}{k}$

9a. $b_2 = a_1 + a_2$

9b. $b_3 = a_2 + a_3$

9c. $b_i = a_{i-1} + a_i$

10. $\binom{n+k}{k} = \binom{m}{k}$

11.

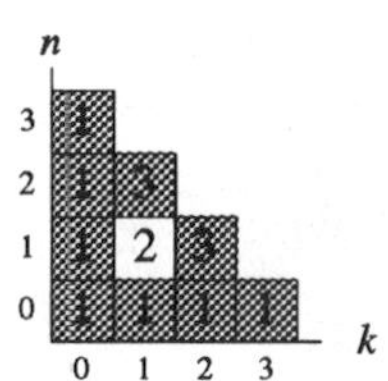

12.

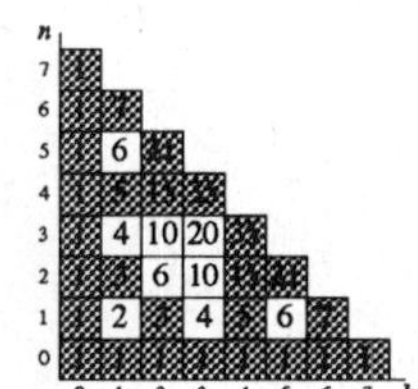

13. 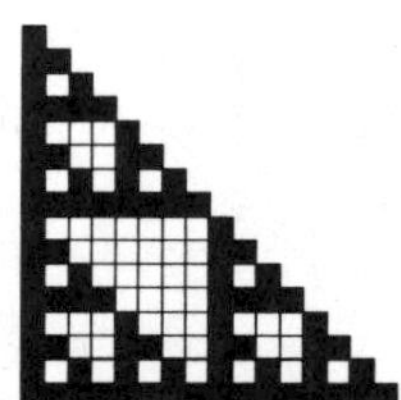

14. The figure in exercise 11 corresponds to stage 2, exercise 12 to stage 3 and exercise 13 to stage 4 . To arrive at the i-th stage of the Sierpinski Triangle you need 2^i rows and columns.

ACTIVITY 7.6

1\.

a). 1001 + 110 = 1110
no carry,
not divisible by 2

b). 10111 + 10000 = 100111
one carry,
divisible by 2

c). 10101 + 1010 = 11111
no carry,
not divisible by 2

2\.

a). 100 + 020 = 120
no carry, not
divisible by 3

b). 212 + 121 = 1110
more than one
carry, divisible by 3

c). 210 + 101 = 1011
one carry,
divisible by 3

3\. 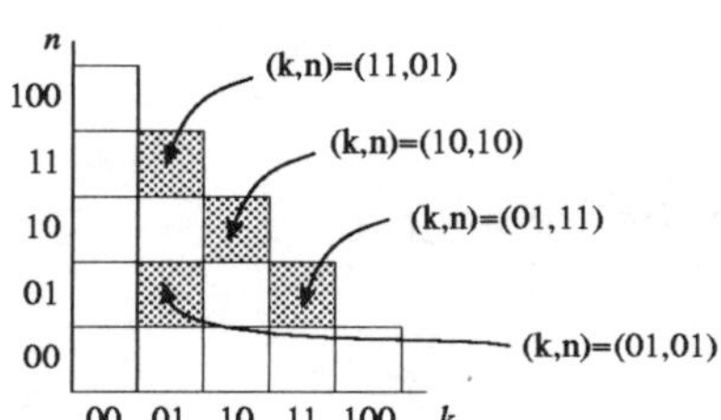

4\. 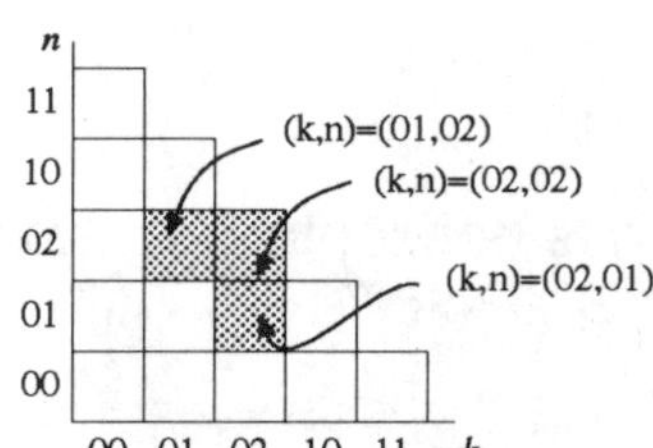

5\. 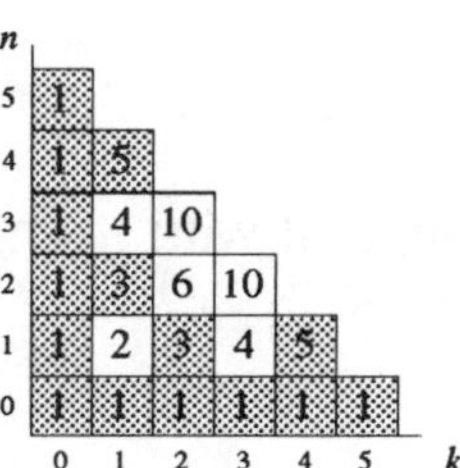

6\. 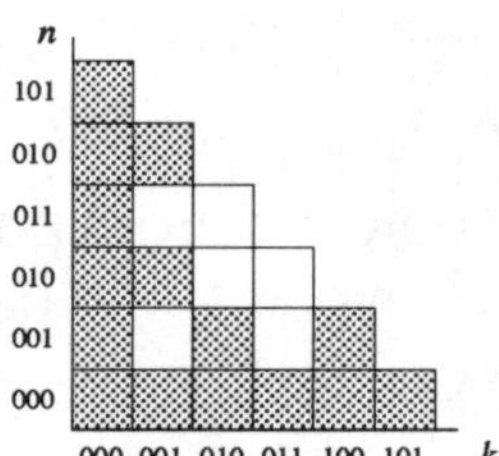

ACTIVITY 7.7

1\. 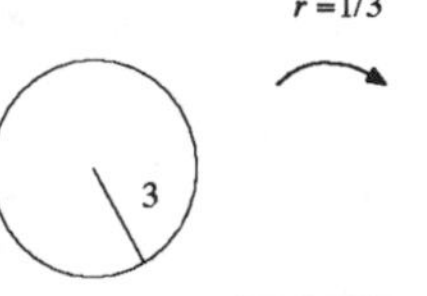

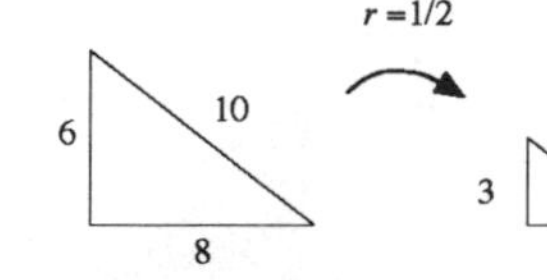

2\. 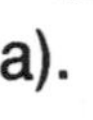 a). 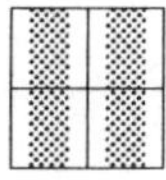 b). 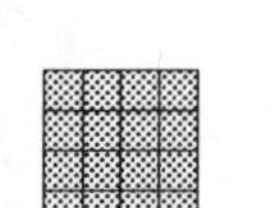 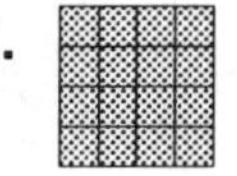 c). 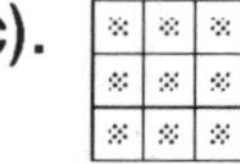

3.

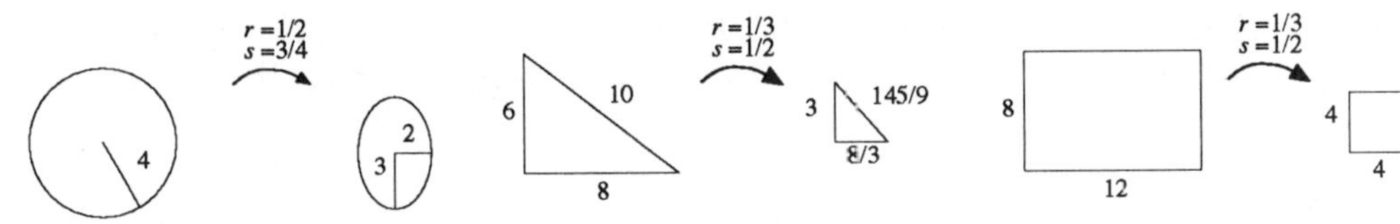

4. Length orig. diagonal: $\sqrt{2}$; Length image. diagonal: $\sqrt{13/16}$
   The multiplying ratio is not equal to r or s. The ratio is $\sqrt{13/32}$.

5. a). 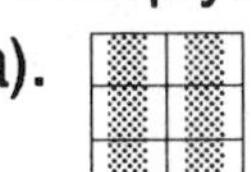 b). 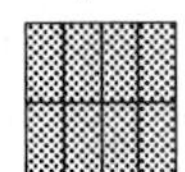 c). 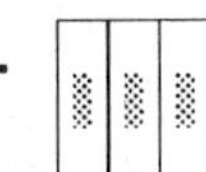

ACTIVITY 7.8

5.

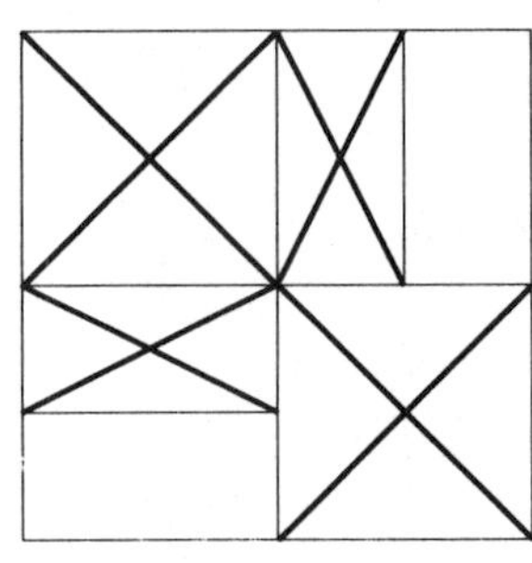
Stage 1

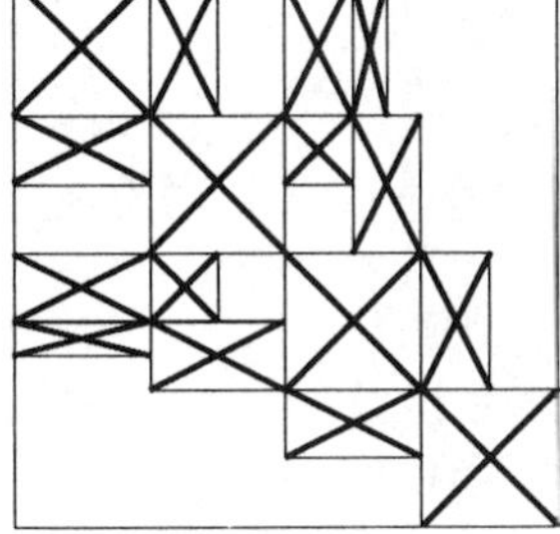
Stage 2

6. No. Because the final image is independent of the initial image, only the transformations are important

7.

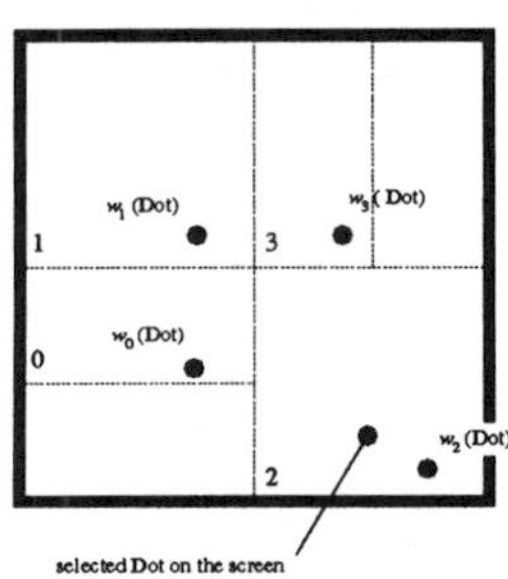

8. The whole image can be partioned into four parts by the four transformations w_0, w_1, w_2, w_3 in such way that each of these parts is a replica of the whole image. Every of these parts again can be partioned into four parts by the four transformations w_0, w_1, w_2, w_3 in such way that each of these smaller parts is a replica of the first partionated parts.
9. Because we know how to build the resulting image. Only these four transformations can build this resulting image.

ACTIVITY 7.9

2. two distinct rotations : 120 and 240 degrees
3. three distinct rotations : 0, 120 and 240 degrees
4. three rotations for each region: $3 * 3 * 3 = 27$

5. When a whole object is self-similar it can be represented by mappings that together comprise the whole object in such way that each image of these mappings is a replica of the whole.
6. 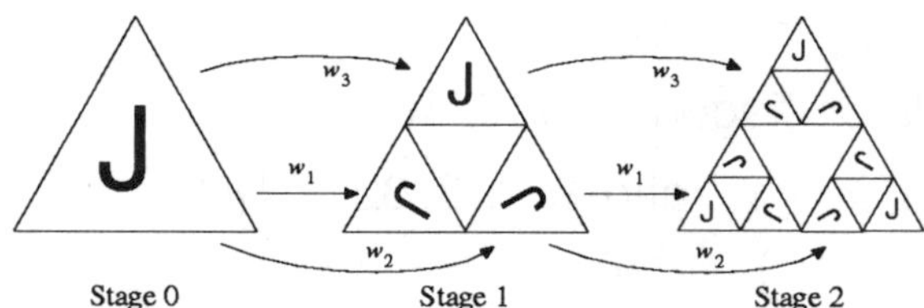

7. No. Because the final image is independent of the initial image, only the transformations are important.
8. 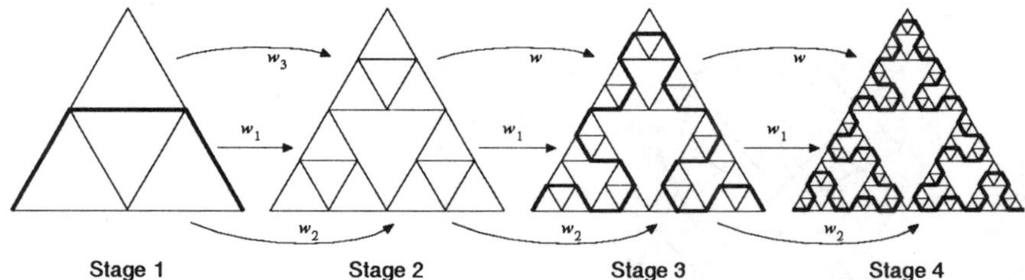

9. The limit figure for the Sierpinski Curve is the Sierpinski Triangle.
10. The Sierpinski Curve is self-similar because it consists of three scaled copies of itself.

UNIT 8 Geometric Genetic Code

ACTIVITY 8.1

1a.

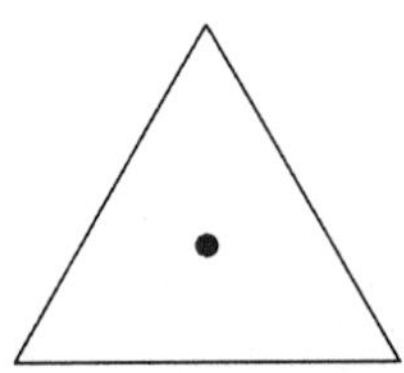

1b.

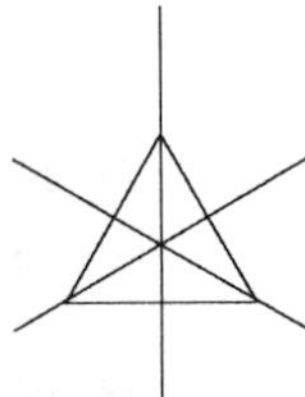

1c. The point of intersection of the lines of symmetry is the rotational point of symmetry.

2.

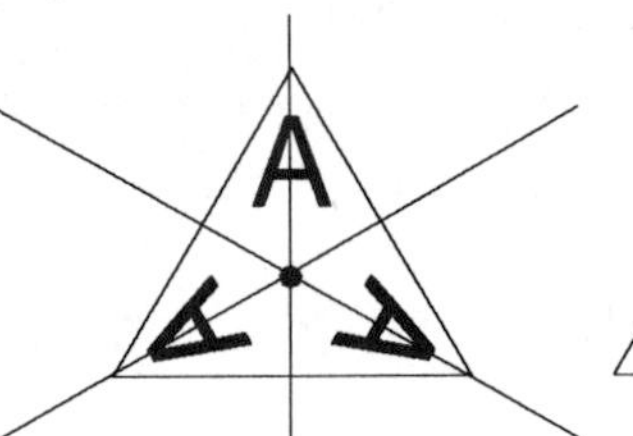
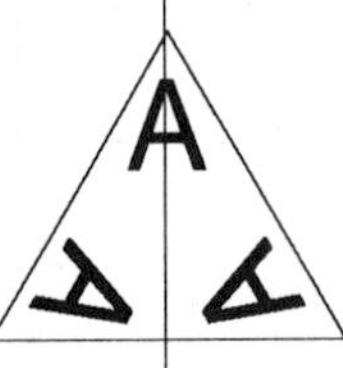
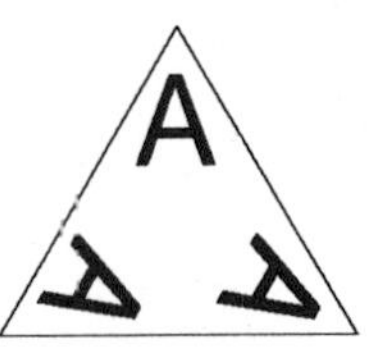
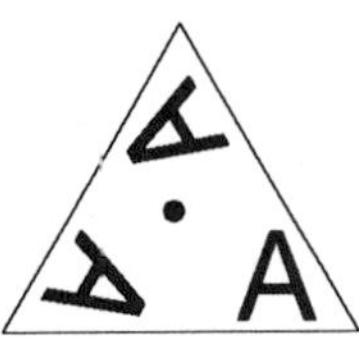

3. A B C D E F G H I J K L M N O P Q R S T U V W X Y Z

3c. F G J K L P Q R Y

4a.

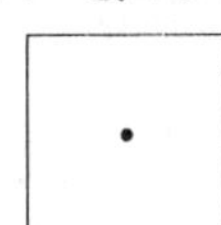

90, 180 and 270 degrees

4b.

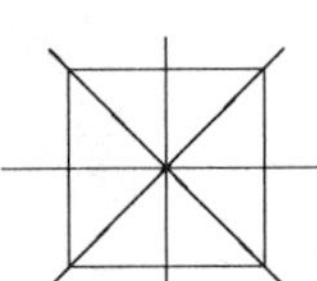

5.

Position 0

Position 1

Position 2

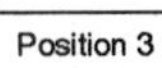

Position 3

Position 4

Position 5

Position 6

Position 7

6.

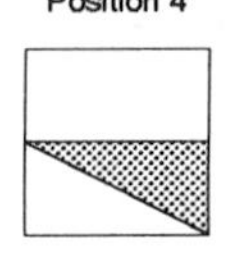

I

R_{90}

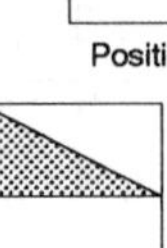

R_{180}

R_{270}

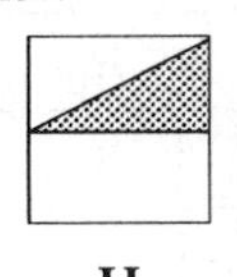

H

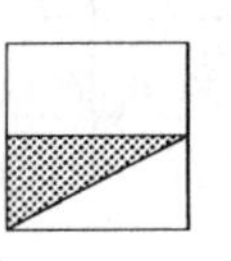

V

D^+

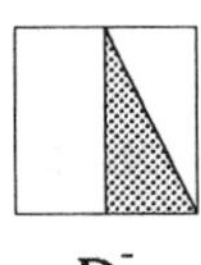

D^-

ACTIVITY 8.2

1. R_{90}
2. D^+
3. R_{180}
4. No, because $R_{270} \circ H = D^-$ and $H \circ R_{270} = D^+$

5. Yes, because $D^- \circ D^+ = R_{180}$ and $D^+ \circ D^- = R_{180}$
6a. V
6b. D^+
6c. R_{270}, Formula : $D^+ \circ V = D^+[V] = R_{270}$
7a. H
7b. D^-
7c. R_{270}, Formula : $D^- \circ H = D^-[H] = R_{270}$
8a. D^+
8b. H
8c. R_{270}, Formula : $H \circ D^+ = H[D^+] = R_{270}$
9a.

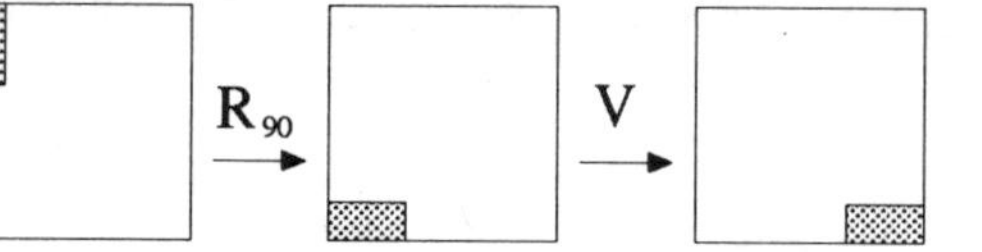

9b.

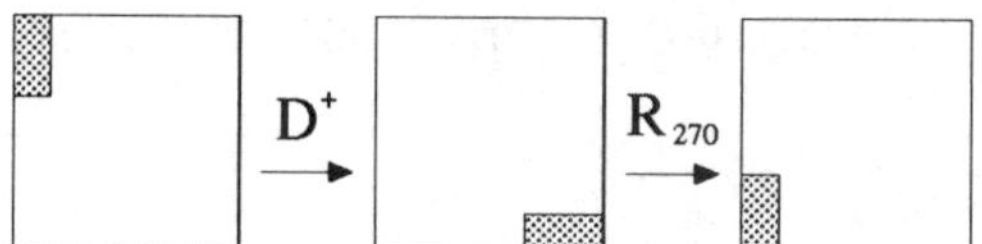

10a. H
10b. H
10c. D^-
11. gives d). because

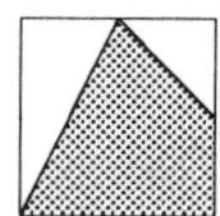

12. gives a). because

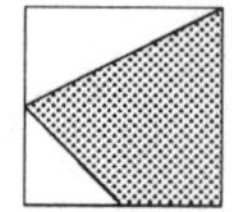
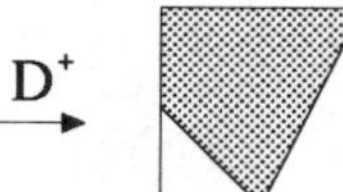

13. gives e). because

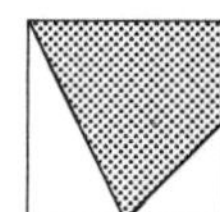
R270 [V]

14. gives a). because

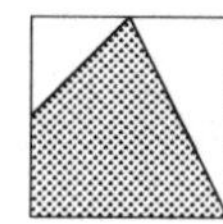
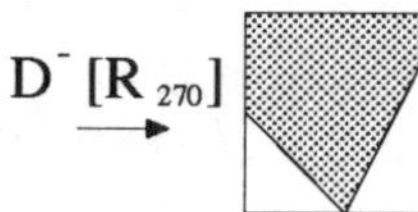

15.

$T_2 \circ T_1$		T_2 = second operation							
		I	R_{90}	R_{180}	R_{270}	H	V	D^+	D^-
	I	I	R_{90}	R_{180}	R_{270}	H	V	D^+	D^-
	R_{90}	R_{90}	R_{180}	R_{270}	I	D^-	D^+	H	V
T_1	R_{180}	R_{180}	R_{270}	I	R_{90}	V	H	D^-	D^+
first	R_{270}	R_{270}	I	R_{90}	R_{180}	D^+	D^-	V	H
oper-	H	H	D^+	V	D^-	I	R_{180}	R_{90}	R_{270}
ation	V	V	D^-	H	D^+	R_{180}	I	R_{270}	R_{90}
	D^+	D^+	V	D^-	H	R_{270}	R_{90}	I	R_{180}
	D^-	D^-	H	D^+	V	R_{90}	R_{270}	R_{180}	I

ACTIVITY 8.3

1.

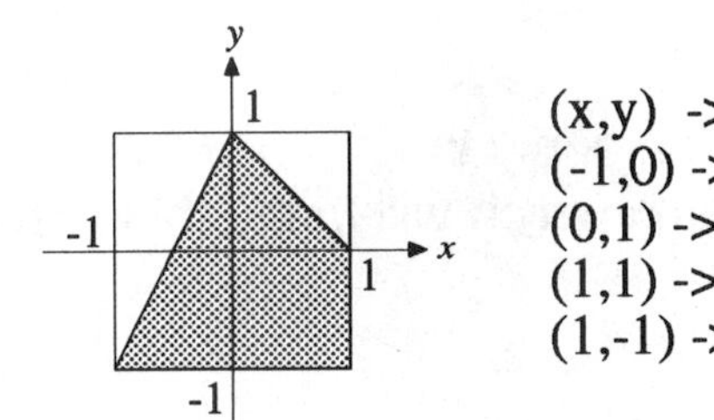

(x,y) -> (x', y')
(-1,0) -> **(0,1)**
(0,1) -> **(1,0)**
(1,1) -> **(1,-1)**
(1,-1) -> **(-1,-1)**

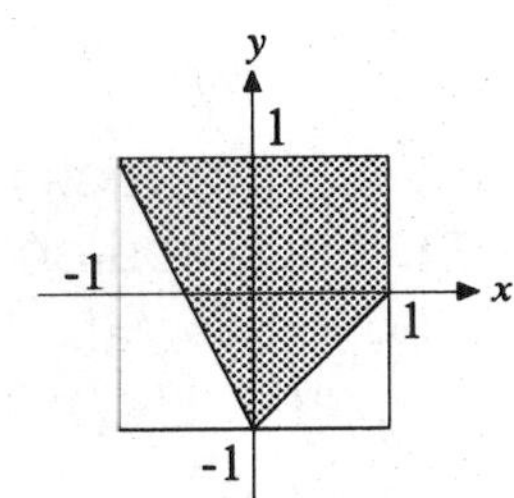

(x,y) -> (x', y')
(-1,0) -> **(0,-1)**
(0,1) -> **(1,0)**
(1,1) -> **(1,1)**
(1,-1) -> **(-1,1)**

2a. (0, 0)
2b. (0, 0.5)
2c. (-0.5, -0.5)
2d. (-0.25, -0.5)
3.

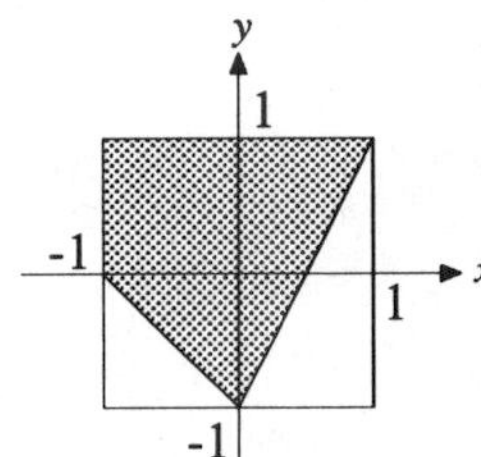

(x,y) -> (x', y')
(-1,0) -> **(0,-1)**
(0,1) -> **(-1,0)**
(1,1) -> **(-1,1)**
(1,-1) -> **(1,1)**

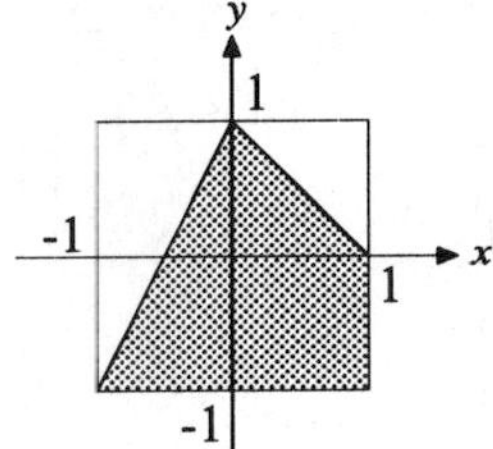

(x,y) -> (x', y')
(-1,0) -> **(0,1)**
(0,1) -> **(1,0)**
(1,1) -> **(1,-1)**
(1,-1) -> **(-1,-1)**

4. Multiply the matrix and the vector.
5a. (1, -1)
5b. (-1, -1)
5c. (1, 1)
5d. (-1, 1)
6a. (1, -1)
6b. (1, 1)
6c. (-1, -1)
6d. (-1, 1)
7. $\begin{pmatrix} 1 & 0 \\ 0 & 1 \end{pmatrix}$
8. Yes. It's equivalent to the reflection about the (main) diagonal. Its represents D^+.
9. (-1, -1), (-1, 1), (1, -1), (1, 1)
10. $\begin{pmatrix} 0 & -1 \\ 1 & 0 \end{pmatrix}$
11. $\begin{pmatrix} -1 & 0 \\ 0 & -1 \end{pmatrix}$
12. $\begin{pmatrix} -1 & 0 \\ 0 & 1 \end{pmatrix}$
13. $\begin{pmatrix} -1 & 0 \\ 0 & 1 \end{pmatrix} \begin{pmatrix} -1 \\ 1 \end{pmatrix} = \begin{pmatrix} 1 \\ 1 \end{pmatrix}$, $V(-1,1) = (1,1)$
14. $\begin{pmatrix} 0 & 1 \\ -1 & 0 \end{pmatrix} \begin{pmatrix} -1 \\ 1 \end{pmatrix} = \begin{pmatrix} 1 \\ 1 \end{pmatrix}$, $R_{270}(-1,1) = (1,1)$

15. $R_{270}(V(-1,1)) = (1,-1)$. The single matrix to get the same effect is $\begin{pmatrix} 0 & 1 \\ 1 & 0 \end{pmatrix}$. For $R_{270}(V)$ is the matrix $\begin{pmatrix} 0 & -1 \\ -1 & 0 \end{pmatrix}$.

16. First apply the matrix of V on the vector $(-1,1)$, then apply the matrix of R_{270} on the resulting vector $(1,1)$. Then search for one matrix which will make this resulting vector again to $(-1,1)$ and this is D^+.
$V \circ R_{270}(x,y) = V[R_{270}(x,y)] = D^-(x,y)$
$$\begin{pmatrix} x'' \\ y'' \end{pmatrix} = \begin{pmatrix} -1 & 0 \\ 0 & 1 \end{pmatrix} \left\{ \begin{pmatrix} 0 & 1 \\ -1 & 0 \end{pmatrix} \begin{pmatrix} x \\ y \end{pmatrix} \right\} = \begin{pmatrix} -1 & 0 \\ 0 & 1 \end{pmatrix} \begin{pmatrix} y \\ -x \end{pmatrix} =$$
$$\begin{pmatrix} -y \\ -x \end{pmatrix} = \begin{pmatrix} 0 & -1 \\ -1 & 0 \end{pmatrix} \begin{pmatrix} x \\ y \end{pmatrix}$$

17a. $R_{90} \circ H(x,y) = R_{90}[H(x,y)] = D^+(x,y)$
$$\begin{pmatrix} x'' \\ y'' \end{pmatrix} = \begin{pmatrix} 0 & -1 \\ 1 & 0 \end{pmatrix} \left\{ \begin{pmatrix} 1 & 0 \\ 0 & -1 \end{pmatrix} \begin{pmatrix} x \\ y \end{pmatrix} \right\} = \begin{pmatrix} 0 & -1 \\ 1 & 0 \end{pmatrix} \begin{pmatrix} x \\ -y \end{pmatrix} =$$
$$\begin{pmatrix} y \\ x \end{pmatrix} = \begin{pmatrix} 0 & 1 \\ 1 & 0 \end{pmatrix} \begin{pmatrix} x \\ y \end{pmatrix}$$

17b. $D^- \circ V(x,y) = D^-[V(x,y)] = R_{90}(x,y)$
$$\begin{pmatrix} x'' \\ y'' \end{pmatrix} = \begin{pmatrix} 0 & -1 \\ -1 & 0 \end{pmatrix} \left\{ \begin{pmatrix} -1 & 0 \\ 0 & 1 \end{pmatrix} \begin{pmatrix} x \\ y \end{pmatrix} \right\} = \begin{pmatrix} 0 & -1 \\ -1 & 0 \end{pmatrix} \begin{pmatrix} -x \\ y \end{pmatrix} =$$
$$\begin{pmatrix} -y \\ x \end{pmatrix} = \begin{pmatrix} 0 & -1 \\ 1 & 0 \end{pmatrix} \begin{pmatrix} x \\ y \end{pmatrix}$$

17c. $D^+ \circ R_{180}(x,y) = D^+[R_{180}(x,y)] = D^-(x,y)$
$$\begin{pmatrix} x'' \\ y'' \end{pmatrix} = \begin{pmatrix} 0 & 1 \\ 1 & 0 \end{pmatrix} \left\{ \begin{pmatrix} -1 & 0 \\ 0 & -1 \end{pmatrix} \begin{pmatrix} x \\ y \end{pmatrix} \right\} = \begin{pmatrix} 0 & 1 \\ 1 & 0 \end{pmatrix} \begin{pmatrix} -x \\ -y \end{pmatrix} =$$
$$\begin{pmatrix} -y \\ -x \end{pmatrix} = \begin{pmatrix} 0 & -1 \\ -1 & 0 \end{pmatrix} \begin{pmatrix} x \\ y \end{pmatrix}$$

ACTIVITY 8.4

0.

A $\mathbf{R_{90}}$	D
B $\mathbf{D^-}$	C $\mathbf{R_{180}}$

1.

R_{270}	
I	I

2.

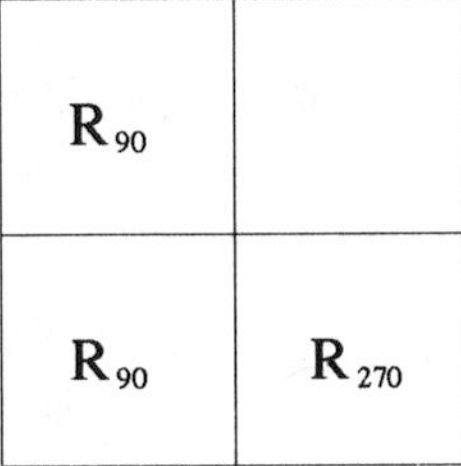

3.

R_{270}	
R_{180}	D^-

4.

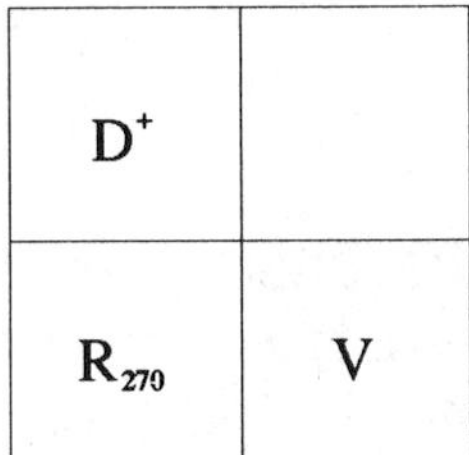

5. The four codes are:
   for question 1 : (R_{270}, I, I)
   for question 2 : $(R_{90}, R_{90}, R_{270})$
   for question 3 : (R_{270}, R_{180}, D^-)
   for question 4 : (D^+, R_{270}, V)
6. (I, H, H)
7. (V, I, I)
8. (I, V, I)
9. (I, H, I)
10. D^+
11. D^+
12. a. (H, I, H) b. (H, R_{270}, H) c. (H, H, H) d. (H, D^-, H)
    e. (H, R_{180}, H) f. (H, V, H) g. (H, D^+, H) h. (H, R_{90}, H)
13.

	a	b	c	d	e	f	g	h
connected	x	x	x	x	x		x	
loops				x	x			

14. a. (R_{270}, R_{270}, V) b. (R_{270}, V, R_{270}) c. (V, R_{270}, R_{270})
15. For example, the two codes (I, I, I) and (D^+, D^+, D^+) both produce the Sierpinski triangle.

ACTIVITY 8.5

1. Cell A : D^-, Cell B : R_{90}, Cell C : V
2. (D^-, R_{270}, V)
3.
4. At stage 3 there are 27. P's. At stage 10 there are $3^{1}0 = 59049$ P's. At stage n there are 3^n P's.
5. The positions of the letters at every stage are important because otherwise one would apply different transformations. The letter itself is not important. There could be anything as initial image because after a few stages the initial image will be a single point. And then there is no significant difference between a letter and another inital image.
6.

H	
H	H

H	
H	V

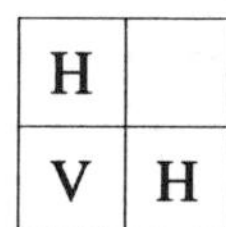
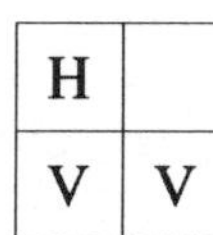
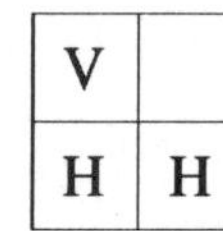
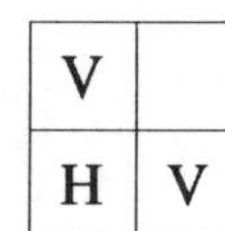
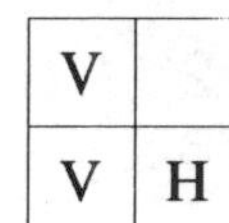
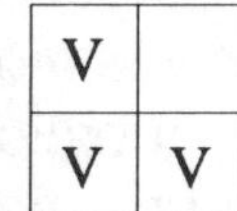

7. $4^3 = 64$ choices
8. $8^3 = 512$ choices
9.

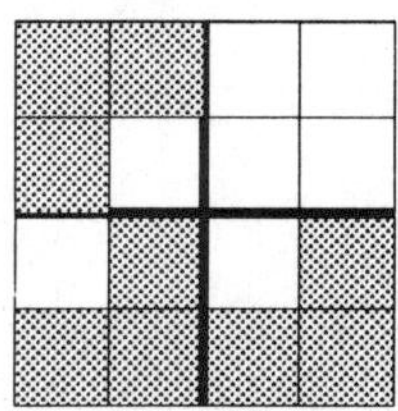
Stage 2

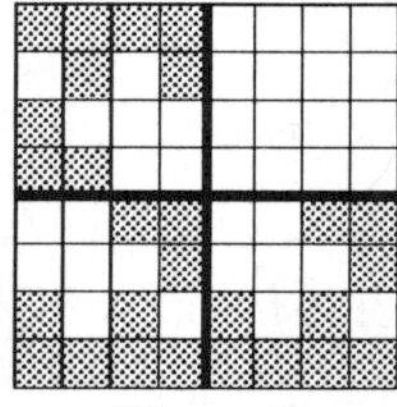
Stage 3

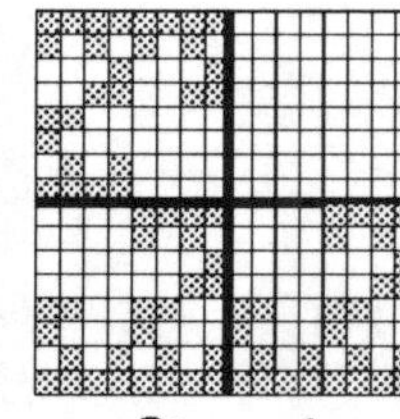
Stage 4

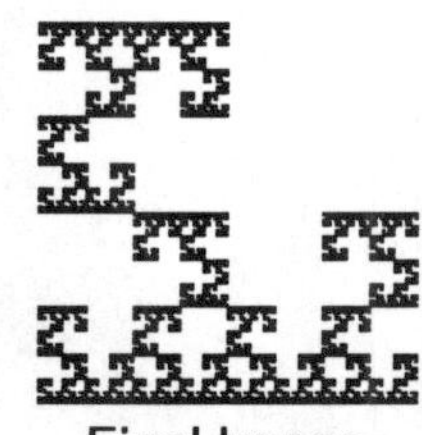
Final Image

10. a). (H, H, H) b). (H, H, V) c). (H, V, H) d). (V, H, H)

    e). (H, V, V) f). (V, H, V) g). (V, V, H) h). (V, V, V)

    Number e). matches the image created in question 9.
11. (D^+, I, D^+), (D^+, D^+, I), (D^+, I, I),
    (I, D^+, D^+), (I, D^+, I), (I, I, D^+)
12.

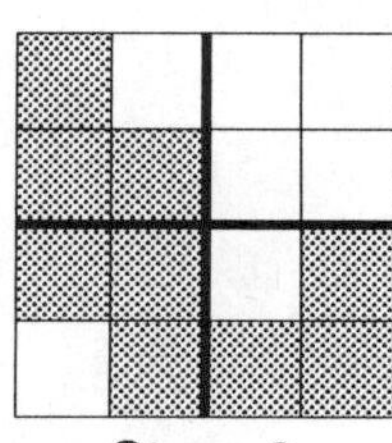
Stage 2

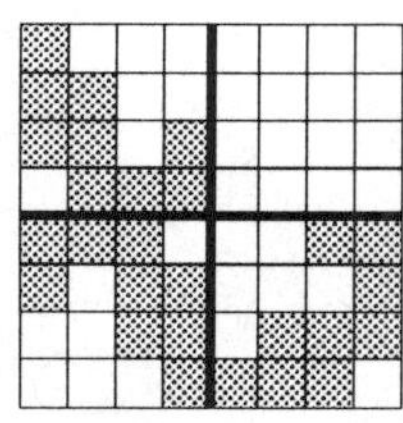
Stage 3

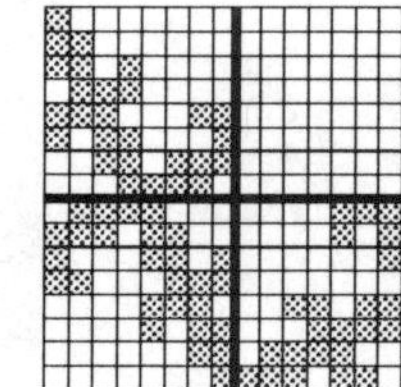
Stage 4

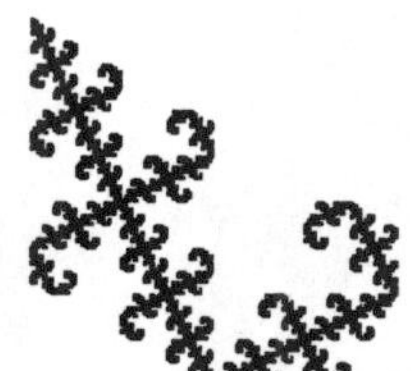
Final Image

16. There would be $3^7 = 2187$ stage-4-squares needed to construct stage 10.

ACTIVITY 8.6

1. The base 2 grid is used in both cases.
2. Three.
3. 1/2.
4. The left fractal (Sierpinski triangle).
5.

Base	Copy	Orient-ation	Scale	Offset Hori	Offset Vert
2	1	I	1/2	-1/2	1/2
	2	I	1/2	-1/2	-1/2
	3	I	1/2	1/2	-1/2

6. I, R_{180}, R_{90}
7. It does match.
8.

Stage 1

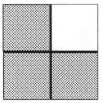

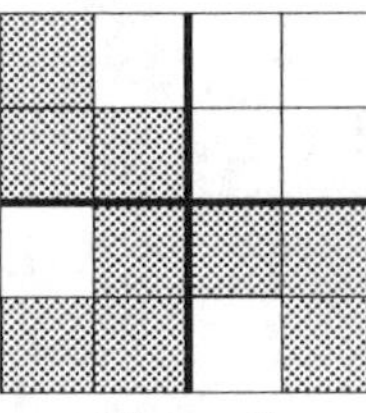
Stage 2

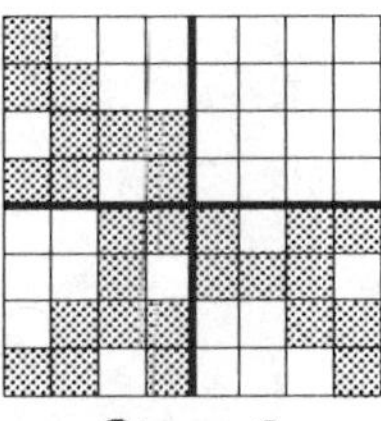
Stage 3

9.

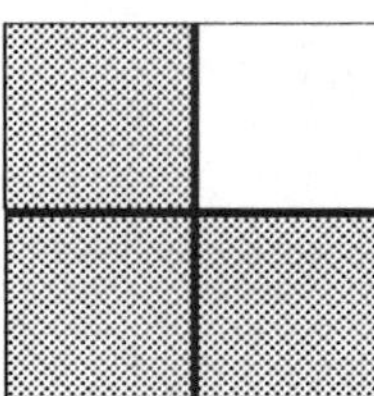
Stage 1

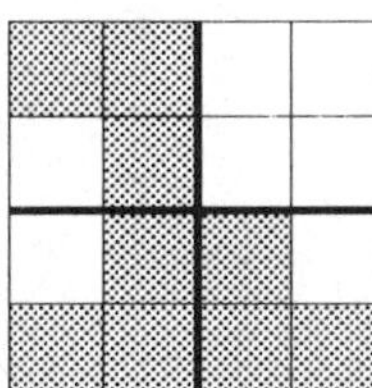
Stage 2

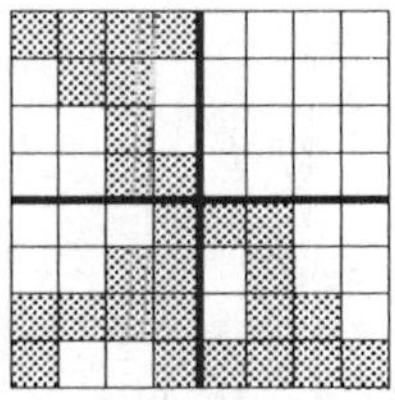
Stage 3

10. The are $3 \cdot 2 \cdot 1 = 6$ different assignments. The resulting fractals are all different (think about the location of the empty subcell in each of the 3 occupied cells).
11. Code for 11:

Base	Copy	Orient-ation	Scale	Offset Hori	Offset Vert
2	1	H	1/2	-1/2	1/2
	2	R_{90}	1/2	-1/2	-1/2
	3	D^+	1/2	1/2	-1/2

Code for 14:

Base	Copy	Orient-ation	Scale	Offset Hori	Offset Vert
2	1	R_{180}	1/2	-1/2	1/2
	2	I	1/2	-1/2	-1/2
	3	V	1/2	1/2	-1/2

Code for 12:

Base	Copy	Orient-ation	Scale	Offset Hori	Offset Vert
2	1	R_{270}	1/2	-1/2	1/2
	2	I	1/2	-1/2	-1/2
	3	R_{180}	1/2	1/2	-1/2

Code for 13:

Base	Copy	Orient-ation	Scale	Offset Hori	Offset Vert
	1	I	1/2	-1/2	1/2
2	2	R_{90}	1/2	-1/2	-1/2
	3	H	1/2	1/2	-1/2

15. Copy 1: $\begin{pmatrix} x' \\ y' \end{pmatrix} = \frac{1}{2}\begin{pmatrix} 1 & 0 \\ 0 & -1 \end{pmatrix}\begin{pmatrix} x \\ y \end{pmatrix} + \begin{pmatrix} -1/2 \\ 1/2 \end{pmatrix}$

Copy 2: $\begin{pmatrix} x' \\ y' \end{pmatrix} = \frac{1}{2}\begin{pmatrix} 0 & -1 \\ 1 & 0 \end{pmatrix}\begin{pmatrix} x \\ y \end{pmatrix} + \begin{pmatrix} -1/2 \\ -1/2 \end{pmatrix}$

Copy 3: $\begin{pmatrix} x' \\ y' \end{pmatrix} = \frac{1}{2}\begin{pmatrix} 0 & 1 \\ 1 & 0 \end{pmatrix}\begin{pmatrix} x \\ y \end{pmatrix} + \begin{pmatrix} 1/2 \\ -1/2 \end{pmatrix}$

16. Copy 1: $\begin{pmatrix} x' \\ y' \end{pmatrix} = \frac{1}{2}\begin{pmatrix} -1 & 0 \\ 0 & -1 \end{pmatrix}\begin{pmatrix} x \\ y \end{pmatrix} + \begin{pmatrix} -1/2 \\ 1/2 \end{pmatrix}$

Copy 2: $\begin{pmatrix} x' \\ y' \end{pmatrix} = \frac{1}{2}\begin{pmatrix} 1 & 0 \\ 0 & 1 \end{pmatrix}\begin{pmatrix} x \\ y \end{pmatrix} + \begin{pmatrix} -1/2 \\ -1/2 \end{pmatrix}$

Copy 3: $\begin{pmatrix} x' \\ y' \end{pmatrix} = \frac{1}{2}\begin{pmatrix} -1 & 0 \\ 0 & 1 \end{pmatrix}\begin{pmatrix} x \\ y \end{pmatrix} + \begin{pmatrix} 1/2 \\ -1/2 \end{pmatrix}$

17.

Base	Copy	Orient-ation	Scale	Offset Hori	Offset Vert
	1	I	1/3	1	1
	2	V	1/3	0	0
3	3	I	1/3	-1	-1
	4	R_{270}	1/3	1	-1

18. Copy 1: $\begin{pmatrix} x' \\ y' \end{pmatrix} = \frac{1}{3}\begin{pmatrix} 1 & 0 \\ 0 & 1 \end{pmatrix}\begin{pmatrix} x \\ y \end{pmatrix} + \begin{pmatrix} 1 \\ 1 \end{pmatrix}$

Copy 2: $\begin{pmatrix} x' \\ y' \end{pmatrix} = \frac{1}{3}\begin{pmatrix} -1 & 0 \\ 0 & 1 \end{pmatrix}\begin{pmatrix} x \\ y \end{pmatrix} + \begin{pmatrix} 0 \\ 0 \end{pmatrix}$

Copy 3: $\begin{pmatrix} x' \\ y' \end{pmatrix} = \frac{1}{3}\begin{pmatrix} 1 & 0 \\ 0 & 1 \end{pmatrix}\begin{pmatrix} x \\ y \end{pmatrix} + \begin{pmatrix} -1 \\ -1 \end{pmatrix}$

Copy 4: $\begin{pmatrix} x' \\ y' \end{pmatrix} = \frac{1}{3}\begin{pmatrix} 0 & 1 \\ -1 & 0 \end{pmatrix}\begin{pmatrix} x \\ y \end{pmatrix} + \begin{pmatrix} 1 \\ -1 \end{pmatrix}$

ACTIVITY 8.7

1. Yes, at stage four the general shape begins to appear. At every higher stage than 6 it's very useful to have the chaos game method.
2. $w_1 : \begin{pmatrix} x' \\ y' \end{pmatrix} = \frac{1}{2}\begin{pmatrix} -1 & 0 \\ 0 & 1 \end{pmatrix}\begin{pmatrix} x \\ y \end{pmatrix} + \begin{pmatrix} -0.5 \\ +0.5 \end{pmatrix}$

   $w_2 : \begin{pmatrix} x' \\ y' \end{pmatrix} = \frac{1}{2}\begin{pmatrix} 0 & -1 \\ 1 & 0 \end{pmatrix}\begin{pmatrix} x \\ y \end{pmatrix} + \begin{pmatrix} -0.5 \\ -0.5 \end{pmatrix}$

   $w_3 : \begin{pmatrix} x' \\ y' \end{pmatrix} = \frac{1}{2}\begin{pmatrix} 0 & -1 \\ -1 & 0 \end{pmatrix}\begin{pmatrix} x \\ y \end{pmatrix} + \begin{pmatrix} +0.5 \\ -0.5 \end{pmatrix}$
3. It's simply some stage in its development because the image is not complete. There are still a lot of points missing. After another 50.000 or more points one can say it's (nearly) the final image.
4. This image is not exactly self-similar. But the three components in cells A, B, C look very close to be replicas of the whole.
5. stage 1 : 32 x 32 Pixels, stage 2 : 16 x 16 Pixels, stage 3 : 8 x 8 Pixels, stage 2 : 4 x 4 Pixels
6. At stage 7 the squares of the grid became smaller than the pixel size. That means that the details on the screen are restricted to the grid size.
9. With random number less than 1/3 the plotted point will be in cell A. With random number greater than 2/3 the plotted point will be in cell C. Otherwise the plotted point will be in cell A.
10. R_{90}
11. D^{-}
12. 
14. In the original program the lines 25 and 26 resp. 37 and 38 should be replaced by the lines 31 and 32 to gain a fractal where the R_{90} mapping is applied to all three cells.
    In the original program the lines 25 and 26 resp. 31 and 32 should be replaced by the lines 37 and 38 to gain a fractal where the D^{-} mapping is applied to all three cells.
16. 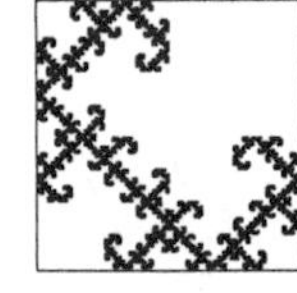
17. 
18. 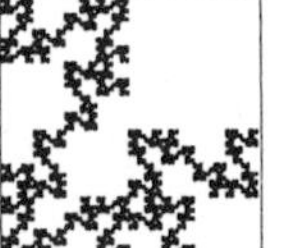
19. (I, R_{90}, R_{180})
20. (I, D^{-}, V)
21. (R_{180}, I, V)

ACTIVITY 8.8

1. (R_{180}, I, D^-), (D^-, I, R_{180})
   (R_{180}, D^+, D^-), (D^-, D^+, R_{180})
2. (V, I, H), (V, I, R_{270}),
   (V, D^+, H), (V, D^+, R_{270}),
   (R_{90}, I, H), (R_{90}, I, R_{270}),
   (R_{90}, D^+, H), (R_{90}, D^+, R_{270})

3a. No line of symmetry, therefore a unique genetic code (I, H, D^-).

3b. $(R_{270}, R_{180}, R_{90})$, (R_{270}, R_{180}, V),
(R_{270}, D^-, R_{90}), (R_{270}, D^-, V),
(H, R_{180}, R_{90}), (H, R_{180}, V),
(H, D^-, R_{90}), (H, D^-, V)

3c. No line of symmetry therefore a unique genetic code (D^+, R_{270}, H).

4.

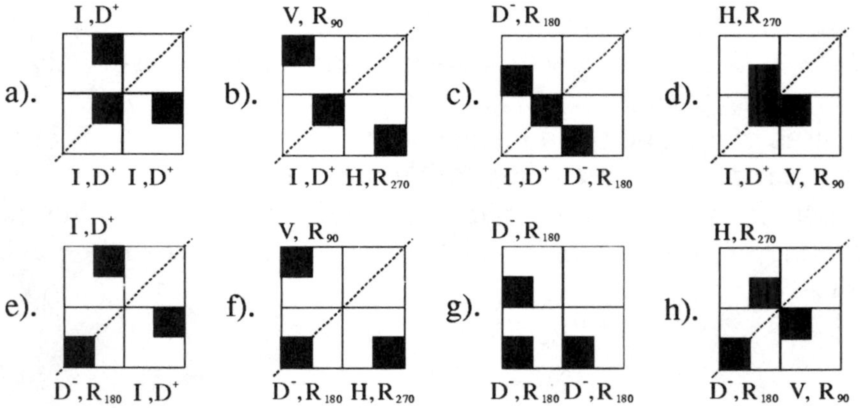

a). I, D^+ / I, D^+ I, D^+

b). V, R_{90} / I, D^+ H, R_{270}

c). D^-, R_{180} / I, D^+ D^-, R_{180}

d). H, R_{270} / I, D^+ V, R_{90}

e). I, D^+ / D^-, R_{180} I, D^+

f). V, R_{90} / D^-, R_{180} H, R_{270}

g). D^-, R_{180} / D^-, R_{180} D^-, R_{180}

h). H, R_{270} / D^-, R_{180} V, R_{90}

5. question 1 : type c, question 2 : type b, question 3 : type h.
6. $512 - 56 = 456$
7. The second is the $D+$ transformed copy of the first one.
8. a. I b. R_{270} c. R_{180} d. R_{90}
   e. H f. V g. D^- h. D^+

	Transformations							
	I	R_{90}	R_{180}	R_{270}	H	V	D^+	D^-
Cell A	P	S	R	Q	Q	S	R	P
Cell B	Q	P	S	R	P	R	Q	S
Cell C	R	Q	P	S	S	Q	P	R
Cell D	S	R	Q	P	R	P	S	Q

10. I and D^+,
R_{90} and V, R_{180} and D^-,
R_{270} and H.

11.

Cell	Original (Codes)	Relative (Codes)
A	R_{270} or H	R_{180} or H
B	R_{180} or D^-	R_{90} or H
C	R_{90} or V	empty
D	empty	R_{270} or V

12.

Cell	Original (Codes)	Relative (Codes)
A	D^-	V
B	R_{90}	empty
C	V	D^-
D	empty	R_{90}

13.

Cell	Original (Codes)	Relative (Codes)
A	R_{270}	empty
B	D^-	R_{180}
C	R_{180}	D^+
D	empty	R_{90}

14.

Cell	Original (Codes)	Relative (Codes)
A	D^+	D^+
B	R_{270}	empty
C	H	V
D	empty	R_{90}

ACTIVITY 8.9

1. Yes

1a. $V \circ R_{90} = D^+$

1b. $D^- \circ D^+ = R_{180}$

1c. $R_{180} \circ H = V$

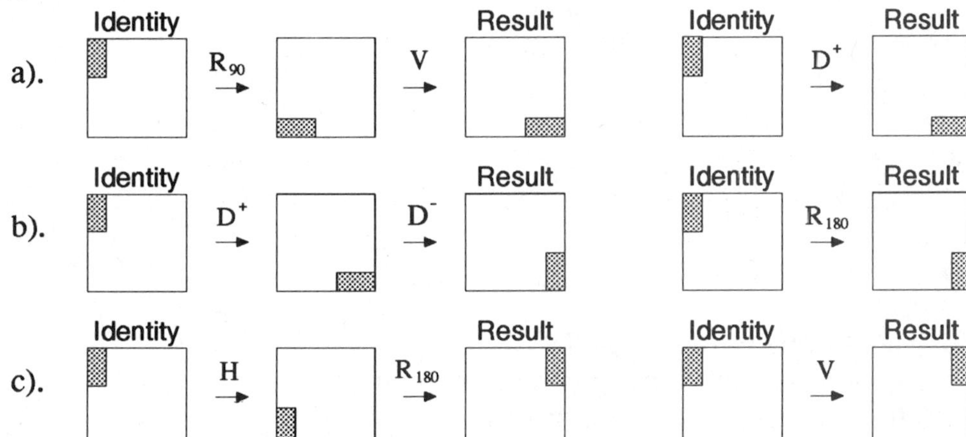

2a. V, R_{180}, I, H

2b. $R_{180}, R_{270}, I, R_{90}$

2c. H, R_{180}, I, V

2d. D^+, R_{180}, I, D^-

3. Yes

4 a. I b. R_{270} c. R_{180} d. R_{90}

e. H f. V g. D^+ h. D^-

5. $R_{90} \circ D^- \circ R_{270} = D^+$

$R_{90} \circ R_{90} \circ R_{270} = R_{90}$

$R_{90} \circ V \circ R_{270} = H$

6.

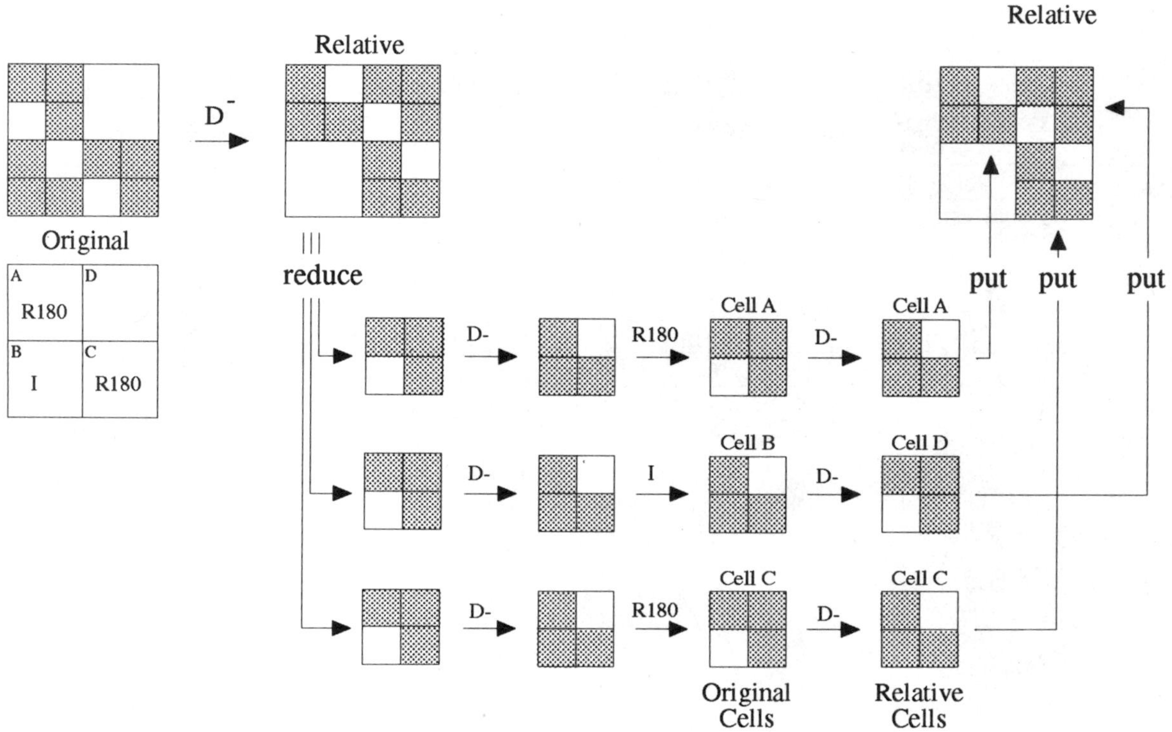

7. $R_{90} \circ I \circ R_{270} = I$
$R_{90} \circ R_{90} \circ R_{270} = R_{90}$
$R_{90} \circ R_{180} \circ R_{270} = R_{180}$
$R_{90} \circ R_{270} \circ R_{270} = R_{270}$
$R_{90} \circ H \circ R_{270} = V$
$R_{90} \circ V \circ R_{270} = H$
$R_{90} \circ D^- \circ R_{270} = D^+$
$R_{90} \circ D^+ \circ R_{270} = D^-$

8a. Motion R_{180}; V is conjugated to itself.
8b. Motion H; R_{90} and R_{270} are conjugates of each other.
8c. Motion D^+; R_{90} and R_{270} are conjugates of each other.
8d. Motion R_{270}; H and V are conjugates of each other.

9.

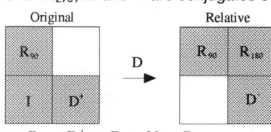

Original: R_{90}, I, D^+ — D → Relative: R_{90}, R_{180}, D^-

$D^- \circ R_{90} \circ D^+ = D^- \circ V = R_{90}$
$D^- \circ I \circ D^+ = D^- \circ {}^+ = R_{180}$
$D^- \circ D^+ \circ D^+ = D^- \circ I = D^-$

10. 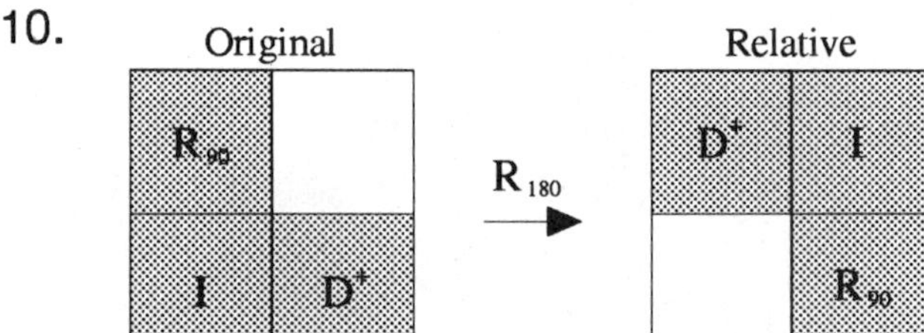

$R_{180} \circ R_{90} \circ R_{180} = R_{180} \circ R_{270} = R_{90}$
$R_{180} \circ I \circ R_{180} = R_{180} \circ R_{180} = I$
$R_{180} \circ D^+ \circ R_{180} = R_{180} \circ D^- = D^+$

11. Cell A : R_{270}, Cell B : R_{180}, Cell C : V

13a. $I \circ R_{90} \circ I = I \circ R_{90} = R_{90}$
$I \circ R_{180} \circ I = I \circ R_{180} = R_{180}$
$I \circ V \circ I = I \circ V = V$

13b. $R_{90} \circ R_{90} \circ R_{270} = R_{90} \circ I = R_{90}$
$R_{90} \circ R_{180} \circ R_{270} = R_{90} \circ R_{90} = R_{180}$
$R_{90} \circ V \circ R_{270} = R_{90} \circ D^- = H$

13c. $R_{180} \circ R_{90} \circ R_{180} = R_{180} \circ R_{270} = R_{90}$
$R_{180} \circ R_{180} \circ R_{180} = R_{180} \circ I = R_{180}$
$R_{180} \circ V \circ R_{180} = R_{180} \circ H = V$

13d. $R_{270} \circ R_{90} \circ R_{90} = R_{270} \circ R_{180} = R_{90}$
$R_{270} \circ R_{180} \circ R_{90} = R_{270} \circ R_{270} = R_{180}$
$R_{270} \circ V \circ R_{90} = R_{270} \circ D^+ = H$

13e. $H \circ R_{90} \circ H = H \circ D^+ = R_{270}$
$H \circ R_{180} \circ H = H \circ V = R_{180}$
$H \circ V \circ H = H \circ R_{180} = V$

13f. $V \circ R_{90} \circ V = V \circ D^- = R_{270}$
$V \circ R_{180} \circ V = V \circ H = R_{180}$
$V \circ V \circ V = V \circ I = V$

13g. $D^- \circ R_{90} \circ D^- = D^- \circ H = R_{270}$
$D^- \circ R_{180} \circ D^- = D^- \circ D^+ = R_{180}$
$D^- \circ V \circ D^- = D^- \circ R_{270} = H$

13h. $D^+ \circ R_{90} \circ D^+ = D^+ \circ V = R_{270}$
$D^+ \circ R_{180} \circ D^+ = D^+ \circ D^- = R_{180}$
$D^+ \circ V \circ D^+ = D^+ \circ R_{90} = H$